# Statistical Sampling for Audit and Control

# Statistical Sampling for Audit and Control

T. W. McRae

*Professor of Finance,*
*The University of Bradford Management Centre*

*A Wiley–Interscience Publication*



John Wiley & Sons
London · New York · Sydney · Toronto

***Library of Congress Cataloging in Publication Data:***

McRae, Thomas W.
Statistical sampling for audit and control.
"A Wiley–Interscience publication."

Bibliography: p.

1. Auditing. 2. Sampling (Statistics)
3. Accounting. I. Title.

HF5667.M18 657'.45 73–19329

ISBN 0 471 58991 8

Printed in Great Britain
by William Clowes & Sons, Limited
London, Beccles and Colchester

*To Cathleen*

# Preface

In recent years the accounting profession has resembled one of the 'forward' battalions in the front trenches during the first world war. The enemy fire has descended so fast and furious that the officers, let alone the foot soldiers, have had little chance to pop their heads above the parapit to see what is going on.

A good deal of the current criticism of accountants and accounting is misguided, but there can be no denying that the 'image' of the profession has taken a severe bruising.

The gravamen of the charge seems to be that traditional accounting methods are old-fashioned and have failed to incorporate the new concepts and techniques of management science. In a word, accounting is not scientific.

I doubt whether accounting ever will be or ever could be 'scientific' in any rigorous sense, although I would rate its chances rather higher than economics or sociology. If, however, a scientific approach to an accounting problem is available the accountant should surely use it.

The application of statistical sampling methods to audit practice is a case in point. If an auditor can replace intuition, hunch, guesswork, etc. with precise numerical estimates then surely he should do so, unless very strong arguments are forthcoming against it.

Statistical sampling is the scientific approach to auditing. By using the method the auditor improves both the accuracy and interest of his work and the image of his profession.

Every empirical study I know which has compared statistical to conventional audit sampling has favoured the former. Almost every criticism of statistical sampling would apply with even greater force to conventional sampling methods, although the critics invariably overlook this point: they compare statistical sampling to an ideal sampling system that never has and never will exist.

Auditing is what economists call a 'labour intensive' activity. A large part of the cost of auditing consists of the cost of skilled labour. This fact is unavoidable considering the nature of the activity.

Professional accounting firms, caught between the Scylla of rising wages and the Charybdis of inelastic audit fees have been forced to reduce the size of samples with which they test the validity of the accounting system they are auditing.

The smaller samples have been justified by two auditing innovations, *depth auditing* and *statistical sampling*.

Depth auditing claims to test the system rather than the individual accounting population. Depth auditing is theoretically an admirable method of auditing. But when one considers (a) the small number of items tested, (b) the linked nature of the items tested, i.e. all of the calculations on one payslip, and (c) the time consuming nature of this type of test, it is easy to show statistically that the auditor's level of confidence in the correct operation of the system must be rather low.

Statistical sampling, on the other hand, provides the *minimum* sample size needed to meet the auditor's requirements. The auditor knows exactly where he is going and the minimum work necessary to get there.

One empirical study (Aly and Duboff, 1971) showed that 'the size of the statistically determined audit sample was smaller by 30% than the personal judgement samples obtained from 158 professional auditors.' (Page 128.)

Some other advantages which auditors can derive from using statistical sampling are as follows.

1. The auditor gains a better understanding of the nature of sampling. For example he learns that the accuracy of prediction from a sample depends upon the absolute size of the sample, not upon the fraction of the population included in the sample.
2. The auditor obtains more precise information about the characteristics of the job he is auditing. He measures the size and dispersion of the values in the population, etc.
3. Unintended bias in selecting a sample to be audited is removed by using random sample methods.
4. The method forces an auditor to state the precise objectives of the audit, the method of test to be employed and the level of confidence he requires in his conclusions.
5. In the writer's experience the more precise deliniation of methods and conclusions makes auditing a more interesting activity. This fact may help towards improving the recruitment of auditors.

So far as I can make out the major argument against statistical sampling is the high set-up and learning cost involved. I hope that this book may help to ameliorate these costs somewhat.

This book is written by one who has used and advised on the application of statistical sampling for a good many years. The book emphasizes practical application rather than theory. As such I hope that it will complement rather than supplant the several excellent textbooks already published in this field.

Confidence in using statistical sampling can only be built up by practice. To this end I have devised the Statistical Sampling Workshop which is published in tandem with this volume.[1]

Ilkley. January 1974

T. W. McRae

[1] This is available from the author, c/o the University of Bradford Management Centre, Emm Lane, Bradford.

# Acknowledgements

I would like to express my thanks to the many people who gave me the benefit of their experience in installing and operating systems of statistical sampling. I am particularly grateful to Professor Herbert Arkin and Kenneth Stringer who found time off from their heavy workloads to grant me lengthy interviews.

I would also like to thank Mr. J. M. Cameron, Mr. Paul Peach, the Edwards and Broughton Publishing Company and the McGraw–Hill Publishing Company for permission to incorporate certain tables originally published by them.

T. W. M.

# Contents

# I

# The Uses of Statistical Sampling in Auditing

## INTRODUCTION

Statistical sampling methods can be used by accountants and auditors for estimating a *proportion* or estimating a *value*. Since evaluation is one of the key objectives in accounting, the latter technique is particularly useful.

A wide range of applications are described in the literature but the only true constraint on its application is the imagination, or rather lack of imagination, of the accountant or auditor. Few accounting tasks are not based on some kind of measurement of a population, and where we measure a population we can sample.

We use statistical sampling either (a) because it is *cheaper* than alternative methods or (b) because it provides a more *accurate* answer to a query. Often it does both.

It is not necessary for an accountant to be expert in the mathematical theory of sampling before he uses statistical sampling methods. Tables are available[1] which provide the required sample sizes once the parameters of the problem are known. However an accountant should be familiar with the *basic concepts* of sampling theory, a much less rigorous requirement. A careful reading of Hoel (1960) will provide a sufficient grounding to meet the latter requirements. Cochran (1963) provides a thorough grounding in the former.

The following chapter will describe some of the applications of statistical sampling in auditing and accounting. First we shall examine those applications which attempt to measure a proportion, such as the proportion of erroneous units in a population, later we shall examine the application of scientific sampling methods to evaluating populations such as inventory, error and bad debts. But first let us take a brief look at the relationship between the five basic sampling methods which are used in auditing and accounting.

[1] See Arkin (1963), Brown and Vance (1961), U.S. Air Force Publication (1960).

## THE FIVE SAMPLING METHODS

Figure 1.1 illustrates the relationships between the five sampling methods used in auditing and accounting.

First we need to differentiate between judgement sampling, when the sampler decides, *in advance*, on the attributes which will determine whether or not the unit is included in the sample, i.e. all debts over £10,000, and random sampling, where every unit of the population has an equal chance of selection on each draw. *Statistical sampling methods of evaluation can only be used when the sample drawn is a random sample.*

The next division is between attribute sampling, where we attempt to estimate the *proportion* of a population with a given condition and variables sampling when we attempt to estimate the *value* of a population.

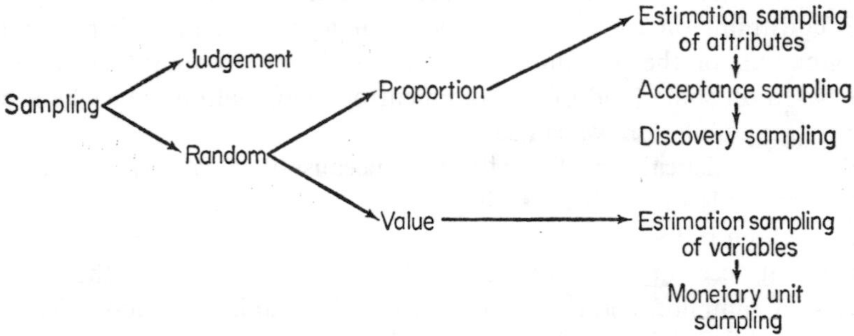


Fig. 1.1 Chart showing relation between sampling methods.

There are three methods of attribute sampling available to us. First *estimation sampling of attributes*, where we estimate the actual proportion of the population having the condition, second *acceptance sampling*, where we test a batch to see whether the proportion of units having the given attribute exceeds a given percentage, while ensuring that we do not reject too many acceptable batches. Third, *discovery sampling*, a simplified form of acceptance sampling which ensures that the proportion of units with a given attribute, say error, does not exceed a given percentage of the population.

Each of these methods requires a smaller sample size than the previous method and consequently each provides rather less information than its predecessor.

For example:

| Population 10,000 | |
|---|---|
| Minimum unacceptable error rate | 2% |
| Expected error rate | 1% |
| Confidence level | 95% |

The required sample sizes are:

| | Sample size |
|---|---|
| (1) Estimation sampling of attributes | 850 |
| (2) Acceptance sampling | 300 (2) |
| (3) Discovery sampling | 150 |

(1) estimates the actual error rate, within an upper and lower bound, (2) checks whether the error rate is less than an upper bound, while seldom rejecting acceptable batches, and (3) checks whether the error rate is less than an upper bound.

We will now examine some practical applications of these three methods.

## DISCOVERY SAMPLING

The simplest method of statistical sampling to apply is discovery sampling. To use discovery sampling we need to know:

1. Population size
2. Minimum unacceptable error rate
3. Required level of confidence in the inference

We then look up suitable sampling tables[1] to find the required sample size. If *none* of the random samples contain an error we conclude that the actual error rate is less than the minimum unacceptable error rate with a given level of confidence.

Discovery sampling is sometimes called exploratory sampling and this latter term provides a clear description of its use. Discovery sampling is seldom used to provide a final answer; it is normally used to pick out batches of documents, etc., which might require more detailed checking. Suppose for example that we wish to check the accuracy of costing of documents arriving from fifteen branches. A small discovery sample check can pick out those batches which have, say, a 95% probability of having an error rate less than 1%. We therefore accept those batches as satisfactory and carry out a detailed check on the remainder.

[1] See Arkin (1963), pp. 525–553.

The method is particularly useful for providing an economical check on the quality of clerical work. One company, for example, uses it to check the accuracy of alphabetic coding in punch cards.

When little time is available for testing the final few months in the year end audit, discovery sampling can provide reassurance that the error rate is below a given percentage with a small sample size.

The method is also useful for testing an audit reliability *expost*. Suppose an error or fraud has been missed by an auditor who has tested the population using a random sample. He can, after the error has been discovered, calculate the probability of his picking up this error. He can state 'I checked $x$ random units and this gave me a, say, 95% level of confidence that the error rate in the population was less than 1%. The erroneous units make up say 0·1% of the population so my method and assumptions are surely adequate. Auditors cannot be expected to find needles in haystacks.'

The weakness of discovery sampling is that it rejects a good number of acceptable batches. This suggests that it is suitable for use as a final check by the internal auditor but should only be used as a preliminary scanning device by the external auditor.

## ACCEPTANCE SAMPLING

We have seen that discovery sampling is a useful method of rejecting unacceptable batches, but that it also rejects too many acceptable batches. Acceptance sampling overcomes this latter defect at the cost of a larger sample size. Acceptance sampling ensures that we do not reject too many good batches, i.e. that we do not throw out too many babies with the bathwater.

Acceptance sampling is particularly useful to the internal auditor who wishes to place a *continuous* control on the quality of clerical work. From suitable tables[1] he can choose a sampling plan which will ensure that errors will not exceed a given percentage of the batch as long as he carries out a complete check of rejected batches.

The internal auditor will have, or ought to have, a diagram of the information channels within the organization. At random times he will inspect the flow of documents along these channels using acceptance sampling. The method can be used to check pricing, coding, arithmetic, authorizations, etc. on the various documents flowing through the information channels.

Acceptance sampling is of rather less value to the external auditor. It

[1] See Arkin (1963), pp. 554–577.

is not easy to find a sampling plan which, while rejecting say 95% of unsatisfactory batches does not also reject too many satisfactory batches. Acceptance sampling is an *internal* audit tool.

## ESTIMATION SAMPLING OF ATTRIBUTES

Discovery and acceptance sampling are used to ensure that the error rate or other proportion of a population does not *exceed* a given percentage. They do not attempt to measure the *actual* proportion of error, etc.

Estimation sampling of attributes attempts to measure the actual proportion of the population having the given condition.

Estimation sampling of attributes can be used in two ways. First it can be used independently to estimate the proportion of, say, incorrectly coded documents, in the population. Secondly, it can be used as a preliminary to employing estimation sampling of variables to estimate the value of a given aspect of a population, say value of debts six months overdue.

There are two approaches to estimation sampling of attributes. The first approach requires the auditor to state the *expected* error rate in the population, the method can test to see if the actual error rate approximates to the expected rate. The best tables available for using this approach are Brown and Vance (1961). The second approach is to take a random sample, measure the proportion of the sample with the given attribute, and calculate from this proportion the probability of the error rate in the population exceeding a given percentage. The best tables available for using this approach are in Arkin (1963), Appendix F. The first method requires a smaller sample size but the second provides a more accurate and useful answer.

There are many applications for estimation sampling of attributes in auditing and accounting. Some of the more common applications are as follows:

Aging of debtors,
Stratification of debtors and inventory,
Stratification of inventory turnover,
Estimation of proportion of back orders,
Estimating error proportion,
Estimating use of resources by departments or cost centres.

The method is of most value where the population to be examined is large, say >10,000, relatively homogeneous, and a precise answer is not required.

In one example known to the author the following conditions applied:

| | |
|---|---|
| Number of customers | 15,000 |
| Estimated proportion of debts three months overdue | 15% |

It was decided to check this estimate using a confidence level of 95% and a precision limit of ±5%. The sample size was found to be 300 and the proportion came out at 37% ±5%. This rough estimate was sufficient to disprove the initial estimate.

Estimation sampling of attributes is frequently used by external auditors to provide a scientific framework for their statements about an accounting population. After estimation sampling of attributes has been used an auditor can make a statement such as follows. 'I am 95% confident that the error rate in this population is less than 0·5%.' However, we must point out that statistical sampling is not much good for finding needles in haystacks. If, for example, only 10 errors occur in a population of 10,000 the probability of picking up *one* error at various sample sizes is as follows:

| Sample size | Probability of picking up one erroneous unit % |
|---|---|
| 100 | 9·6 |
| 500 | 40·1 |
| 1000 | 65·1 |
| 2000 | 89·3 |
| 3000 | 97·2 |

To have even a 90% chance of picking up one erroneous unit the auditor must check the huge sample of 2000 units or 20% of the population. It may be, though, that he can use prior knowledge to *stratify* the population into one group where error is more likely and one where it is less likely, and so considerably reduce the required sample size. Alternatively he could use the MUS system set out in Chapter 17.

## ESTIMATION SAMPLING OF VARIABLES

Attribute sampling estimates proportions, estimation sampling of variables estimates value. The latter technique is far and away the most useful to both the auditor and the accountant. Note, however, that the two techniques are frequently combined as when the auditor must first estimate the proportion of a population having a given condition, i.e. a debt six months overdue, before measuring the total value of this proportion.

When using estimation sampling of variables the auditor must usually estimate the standard deviation of the population.[1] He did not need to do this for estimation sampling of attributes.

The most common use of estimation sampling of variables is to estimate the value of inventory. The Minneapolis Honeywell Corporation, for example, avoided closing down their factory for a week by switching to statistical sampling of inventory. In place of checking 40,000 lots they checked only a random sample of 4000 and later only 806! The value they arrived at came to within eight-tenths of 1% of the value arrived at by a total count. The time required for the count was cut by 60%.[2]

Estimation sampling of variables can replace a perpetual inventory check using a much smaller sample size. Suppose for example that an inventory consists of 12,000 lots and 1000 lots are checked each month in rotation. If the average value of a lot is £100 with a standard deviation of £20 and sampling error of £2, a statistical sample of 370 is sufficient to estimate the *total* value of £1,200,000 at a 95% confidence level.

Alternatively the auditor can continue to use the 1000 rotational check and estimate the total value of inventory to ±£24,000 on £1,200,000 at a 99·9% level of confidence. The latter approach has fitted a scientific framework to the perpetual inventory method.[3]

Estimation sampling of variables can be allied to estimation sampling of attributes to measure inventory turnover. Estimation sampling of attributes estimates the proportion of inventory with a turnover of 1, 3, 6 months, etc. Estimation sampling of variables then evaluates each of these strata within a given precision limit.

Estimation sampling of variables can also be used to estimate the total discrepancy between stock ledger cards and physical count of stock. This differs from the Minneapolis–Honeywell example where only totals were compared.

The evaluation of debtors, and particularly the evaluation of bad and doubtful debts, is another favourite application of estimation sampling of variables. A random selection of debtors can be circularized and the resulting values extrapolated to an estimated total which is compared to the listed total. In the author's experience the problem here is that only 50% to 70% of customers confirm. The auditor can always increase his sample size to allow for this or use negative confirmations[4] but he cannot be certain that the resulting sample is not a biased sample.

[1] See Arkin (1963), p. 108 for an explanation of a simple method of calculating standard deviation. Note that the MUS system avoids this requirement.

[2] See Rudell (1957).

[3] See Hall (1967).

[4] See Davis, Neter and Palmer (1967).

The aging of debtors, is another useful application of estimation sampling of variables.[1] The problem in estimating this kind of value using conventional methods is the heavy cost in clerical labour. Estimation sampling of variables can provide a close estimate at a small fraction of the cost.

For example in one case it took a clerk ten minutes to age one account and there were of the order of 15,000 accounts, i.e. it would have taken at least one man-year of clerical time if only half the account were in arrears! A scientific sample of only 400 accounts in arrears was sufficient to estimate the value of accounts three months in arrears to within ± £40,000 on a total value of £800,000. The method is useful for verifying credit accounts, particularly small value credit accounts, the outstanding value of trading stamps and so on.

Whenever an accounting asset or liability is made up of a large number of units of small value the use of statistical sampling methods should be considered for verifying its value.

Estimation sampling of variables can also be used to evaluate error or wastage. In the case of error, estimation sampling of attributes will have been used to estimate the proportion of the population in error. Estimation sampling of variables will then be used to calculate the total value of error within a given precision limit at the required level of confidence.

Suppose for example that an internal auditor is checking on the accuracy of pricing a large number of low value sale slips. He finds that the error proportion is 5% with a precision limit of ±2%. He next calculates the mean value of the error for pluses and minuses, multiplies these values by his estimate of the number of + and − erroneous units and nets the result to arrive at an estimate of total value of error. Note that if both the estimation sampling of attribute estimate and the estimation sampling of variable estimate are calculated at a 95% level of confidence, the final estimate is only at a 90% level of confidence.

Finally estimation sampling of variables can be used to estimate the proportion of direct revenue or cost which should be allocated between various departments, etc. The best known example of revenue allocation is the division of inter-company debts by the U.S. internal airline services.[2] Each year many passengers who have paid airline A are, at the last moment, diverted to some other airline. One airline alone handles 70,000 flight coupons of this nature per month. A random sample of around 5% estimates the monthly airline debt to within ±1% of the actual value at a 95% confidence level.

[1] See Cyert and Trueblood (1957) page 108.
[2] See Dalleck (1960).

The same principle can be applied to allocating direct cost. The expensive procedure of preparing a detailed cost allocation of the use of some resource between cost heads can be avoided by taking a small sample of the requisition slips.[1] An example would be the allocation of trucking costs between jobs when the trucks make many short trips to several jobs.

## MONETARY UNIT SAMPLING

This is a method of value sampling specifically designed for the *external* auditor. The method was developed by Haskins and Sells, a leading U.S. accounting firm. The MUS system bases its calculation of sample size on all of the populations contained in the trial balance being audited. Thus the sample size is minimized. The method also solves the difficult problem of handling the extreme skewness inherent in most accounting populations.

This is an admirable method of scientific sampling for the external auditor.

## SUMMARY

Statistical sampling provides a scientific framework for auditing and allows us to calculate the minimum sample size needed to support a given inference.

Estimation sampling of attributes, acceptance sampling and discovery sampling can assist an auditor or accountant to estimate a proportion to a given accuracy. Estimation sampling of variables can be used to estimate a value. Monetary unit sampling is suited to the external auditor.

Statistical sampling has been applied to a wide range of applications in auditing and accounting. A cross-section of these applications are described.

## QUESTION SERIES 1

1. Statistical sampling can be used to measure two characteristics of an accounting population. What are these two characteristics?
2. What is the difference between judgement and random sampling?
3. Name three types of attribute sampling. What information does each of these three methods provide?
4. Give an example of the use of discovery sampling.
5. What advantage does acceptance sampling enjoy over discovery sampling? Give an example of the use of acceptance sampling.

[1] See Chapter 14.

6. What are the two approaches to using estimation sampling of attributes to estimate a proportion? (*)[1]
7. Give four accounting applications of estimation sampling of attributes.
8. What *additional* information about an accounting population must an accountant have to use estimation sampling of variables?
9. Give four examples of the use of ES of V.

## SOME ANSWERS TO QUESTION SERIES 1

6. The auditor can
   (a) test a hypothesis about the rate of error in the population,
   (b) estimate confidence limits on the population error rate from the error rate in the sample.

[1] Suggested answers are provided for questions marked with an asterisk.

# 2

# Some Terms Used in Statistical Sampling

## SOME BRIEF DEFINITIONS

Several terms are used in statistical sampling which may be unfamiliar to the reader. Other terms may be familiar but are used in a rather specialized context by the statistician. All of these terms will be defined in some detail later in this book. In this chapter I will present a brief description of several of the more common terms. I hope that these brief definitions might prove helpful to the reader in presenting a quick snapshot of the various factors involved in drawing and analysing a sample from a population.

A *population* or *field* is any group of units with some characteristic in common. A ledgerful of debtors' accounts is an example of an accounting population.

A *unit* is a member of a population. In the previous example each debtors' account is a unit of the debtors' population.

An *attribute* is any characteristic which a unit of a population either possesses or does not possess. For example a debtor is either in arrears with payment or he is not. The characteristic of being in arrears is an attribute of a debtor.

When an attribute is measurable we call it a *variable*. The value of the amount owed by the debtor is measurable therefore it is a variable.

The *parameters* of a population tell us something of the shape and size of the population. Two commonly used parameters are the *mean* or average value of the population, and the *standard deviation* which tells us how the readings in a population are dispersed around the mean value.

Where the readings are not dispersed symmetrically around the mean value we say that the population is *skewed*.

A *stratum* is a section of a population which differs in some respect from the rest of the population.

A *frequency distribution* sets out the actual frequency of occurrence of various states of a variable. A *probability* distribution predicts the likely frequency of occurrence of the various possible states of a variable.

A *sample* is any number of units drawn from a population. A *judgement* sample is a sample when the criteria for including a unit in the sample is decided in advance. A *random* sample is a sample where every unit still remaining in the population has an equal chance of selection on each draw.

The *level of confidence* in an inference from a sample tells us the proportion of times this statement about the population is likely to be true in the long run.

A *confidence interval* tells us the limits of accuracy on an inference. For example 'I am 90% confident that the value of inventory lies between £980,000 and £1,020,000.' The last two figures give us the *lower and upper bound* on the confidence interval of £40,000.

If the mean estimate lies at the centre of the confidence interval we can express the confidence interval as a *precision limit*.[1] In the previous example the precision limit is ± £20,000 on £1,000,000.

Since a sample is unlikely to be a perfect miniature replica of the population from which it is drawn an inference about a population from a sample can be incorrect. This degree of error is called *sampling error*. The degree of sampling error can be measured very precisely if the parameters of the population from which the sample is drawn are known.

*Standard error* is a statistic which allows us to measure sampling error. Roughly speaking it is the standard deviation of a population of sample means.

Each of these terms will be discussed in more detail later in this book.

## SOME TERMS USED IN STATISTICAL SAMPLING

| | |
|---|---|
| Population | Frequency distribution |
| Field | Probability Distribution |
| Unit of Population | Judgement Sample |
| Attribute | Random Sample |
| Variable | Level of Confidence |
| Parameter | Confidence Interval |
| Mean | Lower and Upper Bound |
| Standard Deviation | Precision Limit |
| Skewed | Sampling Error |
| Strata | Standard Error |

[1] Sometimes called *confidence limit*.

# 3

# The Basic Theory of Sampling

## A PRACTICAL EXAMPLE

I was once concerned with the problem of applying costing methods to a large research and development laboratory. The question arose as to the number of jobs which cost less than £500. The R & D manager estimated the proportion to be around 75%.

'No', said the cost accountant, 'I don't agree. I think the proportion is closer to 90%'.

As the matter was of some importance, since it affected the type of costing system to be employed, the cost accountant volunteered to put one of his clerks onto the job of analysing the R & D job cards. One card was kept for each job. There were approximately 1000 job cards.

'But that will take days', said the R & D manager. 'Look, just tell your chap to pull a hundred cards at random and that will give us a near enough answer.' 'How near?' asked the cost accountant. The R & D manager did a ten-second calculation on the back of an envelope. 'Well', he replied, 'If the proportion of the sample comes out around 90% I am 95% sure the actual proportion will not be less than 85%, and if the sample proportion comes out around 75% I am 95% sure that the actual proportion of the population of jobs costing less than £500 will not be greater than 83%. That's to the nearest round figure, of course.'

'But how do I take a random sample?' asked the cost accountant.

'Well 100 is 10% of 1000 and every job is identified with a five digit number, so pull out every job with say a 5 in the junior digit position of the job number. That will give you close to a 10% random sample.'

88 of the 100 job cards sampled turned out to have a cost of less than £500. So we concluded that there was only a 5% chance of the number of jobs in the population costing less than £500 being less than 83%.

The above example presents a simple if rather crude example of the way we can use sampling to provide a quick, cheap and rough estimate of a characteristic of a population.

The key point in the above example is that, having discovered that 88% of the *sample* had the required condition we were able to *infer* that the proportion of the population having the condition would be greater than 83% with a 95% level of confidence.

Anyone can make an inference about a population from a sample drawn from that population. Auditors have done this from time immemorial. But to make an *accurate* inference the sampler needs to have some knowledge of statistical theory.

Let us now pursue the above example a little further. Unless we take a 100% sample there must always be a finite risk that the inference from the sample will be incorrect. This applies to all types of sampling, to the judgement sampling traditionally employed by auditors as well as to scientific sampling. But scientific sampling enjoys an advantage over judgement sampling in that the degree of risk inevitably inherent in making an inference about a population from a sample can be *stated with precision.* This does not apply to judgement sampling methods.

The scientific sampler uses two measures to state the degree of risk he is taking. The first of these is called the *level of confidence*, the second the *precision limit*, sometimes expressed as a *confidence interval.*

## THE LEVEL OF CONFIDENCE

Suppose an internal auditor receives vouchers in batches of 1000 and he wishes to estimate the proportion of vouchers which are coded incorrectly. He draws a pure random sample of 100 vouchers from the batch of 1000. If 10% or 100 of the vouchers are coded incorrectly what is the probability that the auditor will find 10% of the sample incorrect and therefore make the correct inference about the population, i.e. the batch of 1000?

The answer to this question is provided in Table 3.1 which is taken from a table of the binomial distribution. Let us now take the R & D problem and treat it the same way we treated this auditing problem. In the R & D example the sample proportion came out at 88% and the inference was that the actual population proportion was above the lower precision limit of 83% at a 95% level of confidence.

Let us now look at this problem from a slightly different angle. Let us suppose that the *actual* proportion of large jobs (> £500) and small jobs (≤ £500) in the population are as follows:

Large jobs 10%
Small jobs 90%

Thus 100 of 1000 jobs are large jobs.

The clerk selects a random sample of 100 job cards. What proportion of these job cards will consist of *large* jobs?

Since 900 jobs are small jobs it could be that *none* of the job cards will be large jobs. On the other hand there are 100 large jobs in the population so each of the 100 job cards selected *could* be large jobs. Both of these events are highly improbable. Table 3.1 presents the actual probability of various numbers of large job cards turning up.

Table 3.1. Probability of *n* items in sample of 100 having condition if 10% of population of 1000 has condition.

| Number of large jobs detected in sample of 100 | Percentage of times this number of large jobs will be detected in a large number of samples if 10% of population of 1000 jobs are large jobs |
|---|---|
| 0–2 | 00·26 |
| 3 | 00·59 |
| 4 | 01·59 |
| 5 | 03·39 |
| 6 | 05·96 |
| 7 | 08·89 |
| 8 | 11·48 |
| 9 | 13·04 |
| 10 | 13·19 |
| 11 | 11·99 |
| 12 | 09·88 |
| 13 | 07·43 |
| 14 | 05·13 |
| 15 | 03·27 |
| 16 | 01·93 |
| 17 | 01·06 |
| 18 | 00·54 |
| 19 | 00·24 |
| 20 | 00·12 |
| 21–100 | 00·02 |
| | 100·00 |

Since 10% of the population consists of large job cards we should expect intuitively that the most likely situation is that 10% of the sample or 10 job cards would be large jobs. Table 3.1 confirms this assumption. In 13·19% of cases in a very large number of draws of 100 cards 10 large job cards would turn up. But notice that 9 large job cards would occur almost as frequently (13·04%) and 11 occurs quite frequently (11·99%).

Thus we conclude that if $x$% of the population has a given condition and we draw a sample of $y$% from the population only relatively infrequently will we find that the proportion of the sample is *exactly* $x$%. *However the sample proportion will lie close to* $x$% *most of the time.*

Figure 3.1 illustrates this point: the sample proportion could lie anywhere between 0% and 100% but most of the time the sample proportions cluster near to 10%.

We can set this out in the form of a table as follows:

Table 3.2. Confidence interval and confidence level for job card problem

| No. of times this number of large job cards occurs (Confidence interval) | Percentage of times (Confidence level) |
|---|---|
| 10 | 13·19 |
| 9–11 | 38·22 |
| 8–12 | 59·58 |
| 7–13 | 75·90 |
| 6–14 | 86·99 |
| 5–15 | 93·65 |
| 4–16 | 97·17 |
| 3–17 | 98·82 |
| Remainder | 100·00 |

Although 10 large job cards turn up only about 13% of the time, between 5 and 15 large job cards turn up about 93·65% of the time. Thus we can calculate or, if we can find tables of the probability distribution set out in Table 3.2, we can select any level of confidence we desire.

Returning to our example we might decide to test the cost accountant's assumption that the proportion of large jobs is 10%. We draw our random sample of 100 and find 12 large jobs, does this confirm or refute the cost accountant's assumption?

From Table 3.1 we see that if the actual proportion is 10% then 12 large job cards will occur quite often. The difference (12 − 10) = 2 is caused by *sampling error*. That is, is caused by the sample not including 100% of the population. However, if only 10% of the population consists of large job cards, more than 16 or less than 4 large job cards will occur relatively infrequently. Only about 3% of the time according to Table 3.2. Therefore we can state that if 10% of a population of 1000 units has a given condition then if a sample of 100 is drawn from this population the proportion of the sample having the condition will lie between 4 and 16 units. We make this statement with a (100 − 3) = 97% level of confidence.

In the previous discussion we knew that the actual proportion in the population was 10% and we estimated how close the sample proportion was likely to be to 10%. In auditing practice the situation is precisely the reverse. The auditor knows the actual proportion in the sample, 12 out of 100 in the case discussed above, but he does not know the proportion in the population. This is what he wants to find out. He can approach the problem of estimating the proportion in the population in one of two ways.

1. He can use the sample proportion to test a hypothesis that the population proportion is $x\%$.
2. He can use the sample proportion as a best estimate of the population proportion and calculate the standard error of the sample.

Let us examine method (1).

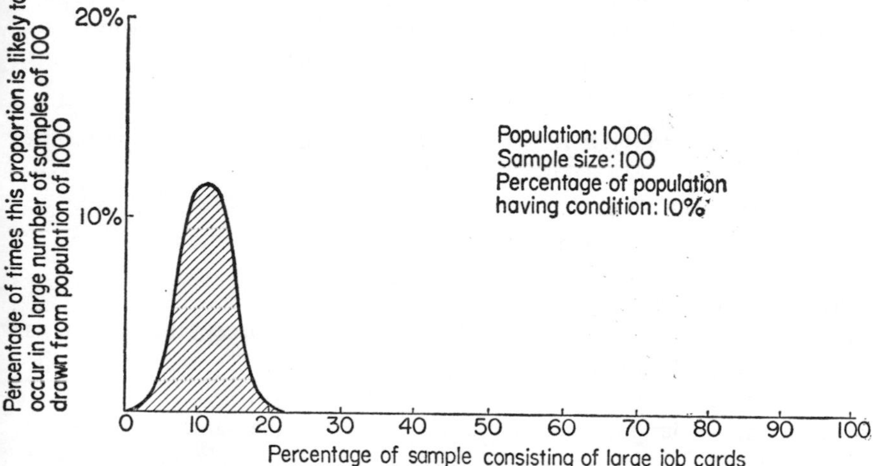


Fig. 3.1. Probability distribution of sample proportions when 10% of population has given condition.

## USING THE SAMPLE TO TEST A HYPOTHESIS

From now on we will call large job cards L and small job cards S.

We recall that the cost accountant put forward the hypothesis that the number of Ls was 10%. A sample of 100 job cards was drawn and 12 Ls were discovered. Did the sample support the cost accountant's hypothesis? This would depend upon the precision limit (confidence interval) and confidence level chosen by the cost accountant *before the sample was drawn.*

The cost accountant might have argued something like this. 'If I draw a sample of 100 from 1000 even if the proportion of Ls in the population is 10% it is rather unlikely that I will find exactly 10% of Ls, i.e. 10 units, in the sample. However, I should find between 5 and 15 units 93·65% of the time; this is pretty close to 95% if the actual proportion in the population is 10%. In other words, if the number of Ls turns out to be between 5 and 15 units I can be almost 95% confident that the population proportion *is* 10% and I will put the difference between 10% and the actual proportion in the sample down to sampling error.'

Using this method, the sampler

1. States a hypothesis about the population proportion before taking the sample.
2. States a confidence level and precision limit for testing the hypothesis.
3. Draws a random sample of given size and counts the proportion in the sample with the given attribute. If this proportion falls within the precision limits stated in (2) he accepts the hypothesis stated in (1) as being true. If the sample proportion falls outside the precision limits he rejects the hypothesis.

Applying this to the R & D job card problem gives the following result.
Hypothesis: The proportion of Ls is 10%.
Confidence level: 93·65%.
Precision limit: 5–15 units.
Size of sample: 100.
Proportion of Ls: 12.
Decision: Since 12 falls in the precision limit 5–15 we accept the hypothesis that the proportion of Ls in the population is 10%. We assume that the difference (12 − 10) = 2 is caused by sampling error.

But, the reader may complain, we could use the same reasoning to justify the R & D manager's estimate of 25% of Ls. The proportion of Ls in the population could be 25% making 25 Ls out of 100 the most likely sample proportion. The actual sample proportion of Ls is 12, but the difference (25 − 12) = 13 can be put down to sampling error?

This is true. It *could* be, but it is rather unlikely. If the population proportion is 25%, the chance of drawing a sample with a proportion of 12% or less is very small, only about one chance in eight hundred in fact. Therefore if we are testing the hypothesis that the population proportion is 25% and our sample of 100 throws up only 12% of Ls we can reject the hypothesis with a level of confidence in excess of 99%.

'Ah, that is all very well,' says the sceptical reader. 'You have convinced me that the proportion of Ls is more likely to be 10% than 25%, but you have not convinced me that *it is actually 10%*!'

I fear there is no answer to this retort. Unless we carry out a 100% check of the population and impose a rigorous audit on the accuracy of the checker we can never be 100% certain of the actual population proportion. And this statement is valid no matter what system of sampling we use.

However, scientific sampling does give us the probability of being wrong! Other sampling methods do not give the auditor this information.

## USING THE SAMPLE PROPORTION AS A BEST ESTIMATE

In the previous section we discussed a situation where two hypotheses about a population proportion were put forward and one of these was accepted and one rejected.

But after reading this section the reader may have thought to himself: 'If the sample proportion is 12% it is pretty obvious that the population proportion is more likely to be 10% than 25% but surely the most likely population proportion is 12%?'

This, again, is quite true. If we have no prior information about the population and we take a random sample of any size from the population the most likely population proportion is the sample proportion.

However in many, if not most, auditing and control applications of sampling the auditor wishes to estimate whether the population exceeds a given percentage or lies between two percentages or is less than a given value or lies between two values. In all of these cases where a given result is expected it is better to tackle the problem by setting up a hypothesis to be accepted or rejected.

If, however, very little prior information is available so that the auditor cannot make an estimate in advance, he will use the sample proportion as a best estimate of the population proportion. He can use the estimate of sampling error, as explained in the following section to place a precision limit and confidence level on his estimate of the population proportion.

For example the cost accountant could calculate that if a sample of 100 is drawn from a population of 1000 and is found to have 12 Ls, then he can state with a 95% level of confidence that the actual proportion of Ls in the population lies between 7% and 19%.

Usually the auditor will use the hypothesis method to make a first attempt at estimating the proportion. Later he will use the sample proportion to check on the likelihood of the population proportion being different from the estimated proportion.

We shall return to this point in a later chapter.

## WHAT SIZE OF SAMPLE?

Up till now we have used a fixed sample size of 100 from a fixed population size of 1000.

In a real audit the population size is given and tends to vary from year to year. It is what statisticians call an uncontrollable variable. The sample size, however, is under the control of the auditor. How large a sample should the auditor draw?

If the reader turns back to Table 3.2 he will see that confidence level can be played off against precision limit at a fixed sample size. When the precision limit is very narrow the confidence level is very low. By widening the precision limit we can increase the level of confidence at a fixed sample size. That is, the bigger the target, the more confidence we have of hitting it.

Later we will see that the standard deviation, i.e. dispersion, of the readings around the mean value, can also affect the sample size.

We conclude that the size of sample drawn by an auditor from a population depends upon two uncontrollable and two controllable factors.

The uncontrolled factors are:

1. Population size.
2. Standard deviation (dispersion) of population.

The controllable factors are:

1. The precision limit.
2. The level of confidence.

In the next chapter we will discuss how an auditor decides on what precision limits and confidence levels to use in a given situation.

## STANDARD ERROR OF THE SAMPLE MEAN

But before we leave this chapter on the basics of sampling we must introduce the concept of *standard error*.

The reader will have noted from the previous discussion that it is not difficult to calculate a sample mean for use as an estimator of the population mean. All the sampler has to do is to take any two units of the population, add them together and divide by two, and he has an estimate of the population mean. But how accurate is this estimate? In other words what degree of sampling error is associated with this estimate of the population mean? With a sample of two from a large population it is likely that the degree of sampling error will be so great that the estimate will have little value as a predictor of the population mean.

The *standard error of a sample estimate* is a statistical device which helps us to measure the degree of sampling error inherent in a sampling estimate.

Let us first examine the standard error of a proportion. This is given by the formula:

$$S = \left(\frac{p(1 - p)}{n}\right)^{1/2}$$

where $S$ = the standard error of the proportion, $p$ = the proportion of the sample having the given condition, $n$ = the number of units in the sample.

This formula assumes that the *sample* makes up a very small proportion of the population, say less than 5% and has an absolute size larger than 30. This is usual in auditing and control applications of sampling. Suppose for example, that we select a random sample of 300 debtor balances from a debtors ledger containing 10,000 balances and we find 30% of them to be three months overdue. What is the standard error of the proportion?

$$S = \left(\frac{30 \times 70}{300}\right)^{1/2} = 2{\cdot}646$$

But how, the reader may ask, does this statistic help us to measure the reliability of the sample proportion as an estimator of the population proportion?

The answer is that we treat the standard error of the sample in the same way that we treat the standard deviation of the population.

It can be proved that the sample estimates of the population mean will usually form a Normal distribution around the population mean. Therefore[1] we can state that the true population mean will lie within:

1 standard error of the sample estimate 68·3% of the time.
2 standard errors of the sample estimate 95·4% of the time.
3 standard errors of the sample estimate 99·7% of the time.

Alternatively we can say that we have a:

90% level of confidence that the true population mean lies 1·68 standard errors from the sample estimate.
95% level of confidence that the true population mean lies 1·96 standard errors from the sample estimate.
99% level of confidence that the true population mean lies 2·57 standard errors from the sample estimate.

In the above problem we can say, for example, that we are 95% confident that the proportion of the population of debtors whose debts are three

[1] See Table 21.1 on p. 238.

months overdue lies between 30% ± (2·646 × 1·96), that is between 24·81% and 35·19%, or to the nearest round figure 25% to 35%.

The precision limits for other levels of confidence are:

| Level of confidence | Precision limits |
|---|---|
| 90% | 25·55–34·45 |
| 99% | 23·20–36·80 |

If the sample makes up 5% or more of the population it is advisable to use the more precise formula

$$S = \left(\frac{p(1-p)}{n}\right)^{1/2}\left(1 - \frac{n}{N}\right)^{1/2}$$

where $N$ is the size of the population. Note that if $n$ is small relative to $N$ the right-hand expression disappears.

When the *proportion being estimated* is very small the above formulae become rather unreliable. For example with proportions of 10%, 5% and 1%, a sample of 300, population size of 10,000 and 95% level of confidence, the precision limits using the above formulae and the true precision limits are:

Table 3.3. Confidence limits on a proportion

| Proportion of sample having condition | Precision limits on estimate of population proportion | |
|---|---|---|
| | Formulae | True |
| 1% | 0·00–2.12 | 0·2–2·9 |
| 5% | 2·53–7·47 | 2·9–8·1 |
| 10% | 6·61–13·39 | 6·9–13·9 |
| 30% | 24·81–35·19 | 24·9–35·4 |

We see that the formula presents a figure rather lower than the true figure for both the upper and lower precision limit, but over about 10% the difference is hardly likely to be significant for accounting and auditing purposes.

When the proportion being investigated is less than 10% readers are advised to use specially calculated tables. Arkin (1963), Appendix F, provides such a table.

Note that the error rates in most accounting populations are likely to be well below 10%, and so if the formula is used the auditor will be *under-*

*estimating* the probability of the error proportion in the population exceeding a given figure.

## THE STANDARD ERROR OF OTHER STATISTICS

The previous section discussed the problem of measuring the standard error of a proportion, and so estimating the sampling error of the population estimate.

The sampling error of any other statistic can also be measured by calculating the standard error of the statistic.

A selection of formulae for measuring the standard error of various statistics is given in Table 3.4.

Table 3.4. Formula for calculating the standard error of some other statistics. $\sigma$ = standard deviation; $N$ = number of units in the population; $\mu_4$ = the fourth moment, i.e. kurtosis; $\mu_2$ = the second moment, i.e. variance, i.e. $(\mathrm{sd})^2$.

| Statistic | Standard error |
|---|---|
| Mean | $\sigma_X = \dfrac{\sigma}{N^{1/2}}$ |
| Standard deviation[a] | (1) $\sigma_s = \dfrac{\sigma}{(2N)^{1/2}}$ <br> (2) $\sigma_s = \left(\dfrac{\mu_4 - \mu_2^{\,2}}{4N\mu_2}\right)^{1/2}$ |
| Variance[a] | (1) $\sigma_{s^2} = \sigma^2\left(\dfrac{2}{N}\right)^{1/2}$ <br> (2) $\sigma_{s^2} = \left(\dfrac{\mu_4 - \mu_2^{\,2}}{N}\right)^{1/2}$ |
| Median | $\sigma_{\text{med}} = \dfrac{1{\cdot}2533\sigma}{N^{1/2}}$ |

[a] If $N < 100$ use formula (2).

## A COMMENT ON THE 'ACTUAL' ERROR RATE

In this book we will frequently use the term 'actual' error rate in the population. By this term we mean the estimate of the error rate we would come up with if we examined the entire population in the way we examined the sample. It is possible that if our method of examining the sample is invalid then even if we examined the entire population using this method we would not arrive at the true proportion of the population having the sought for condition.

This point is sometimes overlooked by auditors.

## QUESTION SERIES 3

1. What major advantage does statistical, i.e. scientific, sampling enjoy over traditional judgement sampling?
2. The sampler states the degree of risk he is willing to take in making an estimate by stating two factors. What are these two factors?
3. Using the data from Table 3.1 on p. 15 calculate the given level of confidence that if a population has 10% of defectives a random sample of 100 units will contain:

   (1) less than 8% defective.
   (2) more than 15% defective.
   (3) between 9% and 11% defective. (*)

4. Using the data in Table 3.1 on p. 15 calculate a confidence interval which provides a 90% level of confidence in the inference. (*)
5. If an auditor draws a sample from a population and finds $x\%$ of the sample in error he can use this information to make an inference about the population in two ways. What are these two ways?
6. State the three steps needed to test a hypothesis about a population containing a given proportion of units with a given condition.
7. If an auditor has no prior information about a population how can he calculate the likelihood of the population having a proportion with a given condition?
8. Write down the formula for finding the standard error of a proportion.
9. What does the standard error of a proportion measure?
10. What level of confidence do we have that the true population proportion lies

    1·68 standard errors from the sample estimate?
    1·96 standard errors from the sample estimate?
    2·57 standard errors from the sample estimate?

11. If the sample makes up a significant proportion of the population a rather more elaborate formula should be used. Write down this formula.
12. When the proportion of a sample having a given condition is very small, say 1%, the formula tends to understate the confidence interval

for the likely population proportion. Why do you think this effect occurs? (*)

13. Calculate the standard error on the following sample proportions. (*)

| | Sample proportion % | Size of sample | Population |
|---|---|---|---|
| (1) | 20 | 500 | 50,000 |
| (2) | 10 | 1000 | 1,000,000 |
| (3) | 50 | 400 | 10,000 |
| (4) | 20 | 100 | 4000 |
| (5)[a] | 15 | 300 | 1000 |
| (6)[a] | 30 | 200 | 600 |
| (7) | 1 | 500 | 10,000 |
| (8)[a] | 2 | 300 | 900 |

[a] More precise formula used.

14. Calculate confidence intervals at

90% level of confidence for (1) and (2),
95% level of confidence for (3) and (4),
99% level of confidence for (5) and (6),

for the data in problem 13. (*)

## SOME ANSWERS TO QUESTION SERIES 3

3. (1) 20·68% (2) 3·91% (3) 38·22%
4. There are many possible answers. 5–16 gives a 90·26% level of confidence.
12. The population is so highly skewed that the population of sample means drawn from the population is not symmetrical. See Chapter 11 for a fuller explanation.

13/14.

| | Standard error | Confidence level % | Confidence interval |
|---|---|---|---|
| (1) | 1·79 | 90 | 17–23 |
| (2) | 0·95 | 90 | 8·4–11·6 |
| (3) | 2·50 | 95 | 45·1–54·9 |
| (4) | 4·00 | 95 | 12·4–27·8 |
| (5) | 1·88[a] | 99 | 25·2–34·8 |
| (6) | 2·93[a] | 99 | 22·5–37·5 |
| (7) | 0·44[a] | | |
| (8) | 0·66 | | |

[a] More precise formula used.

# 4

# What Size of Sample Does the Auditor Need?

## INTRODUCTION

In the previous chapter we discovered that the size of sample depends upon four factors.

1. The size of the population.
2. The level of confidence required in the inference.
3. The width of the precision limit (confidence interval).
4. The standard deviation of the population.

In this chapter we will discuss the effect of (1), (2) and (3) on sample size. The effect of standard deviation on sample size will be discussed in the next chapter.

Population size and standard deviation are given. They are uncontrollable variables. The confidence level and precision limit are decided by the sampler. They are controllable variables.

But before we discuss the effects of these factors on sample size first let us take a brief look at the basic theory of calculating an optimal sample size.

## ON CALCULATING AN OPTIMAL SAMPLE SIZE

The larger the sample the more accurate the inference from the sample is likely to be. However, a law of diminishing returns applies to sampling. After a given point is reached the value of the increment of information derived from the next unit sampled is less than the incremental cost of sampling that unit. At this state we should stop sampling. We have found our optimal sample size.

Figure 4.1 illustrates this point. At $n_0$, where the diminishing incremental value of the $n$th unit sampled just equals the constant incremental cost of drawing that unit, we find the optimal sample size.

However, although it is easy to derive a *theoretical* optimal sample size it is rather more difficult to derive this figure in practice. The incremental cost of drawing an additional unit of the sample is relatively easy to calculate, but how do we calculate the diminishing incremental value? This latter is difficult to calculate precisely.

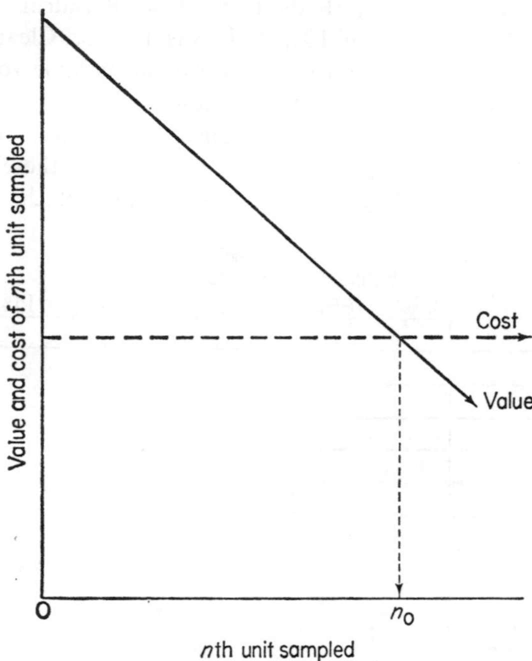


Fig. 4.1. Theoretical solution to deciding an optimal sample size. At that point where the diminishing incremental value of the $n$th unit sampled equals the constant incremental cost of drawing that unit we have our optimal size sample $n_0$.

As we noted above a sampler can calculate a required sample size once he knows:

The population size.
The confidence level required in this inference.
The precision limit required.
The standard deviation of the population.

These factors will generate a given sample size, but they will not tell him if it is an *optimal* sample size. In practice we can only make a rough approximation of the optimal sample size. The auditor can avoid samples which are either much too large or much too small. But he cannot say if the sample is just right.

For example, airlines used to distribute inter-airline ticket debts by randomly selecting a digit between 0 and 9 and drawing all inter-airline passenger flight coupons with this number as its junior digit. The sample was thus always close to 10%. If the number of inter-airline tickets was 10,000 the sample was 1000, if 100,000 it was 10,000. Clearly this was uneconomic. As the population grew the sample became too large. A more sophisticated system is now in operation.

*Depth auditing* is a process whereby an auditor traces one voucher through a set of accounting procedures to test whether these procedures are being operated according to plan. The validity of this method of auditing must be open to doubt since the sample is so small relative to the size of the population of clerical operations being audited.

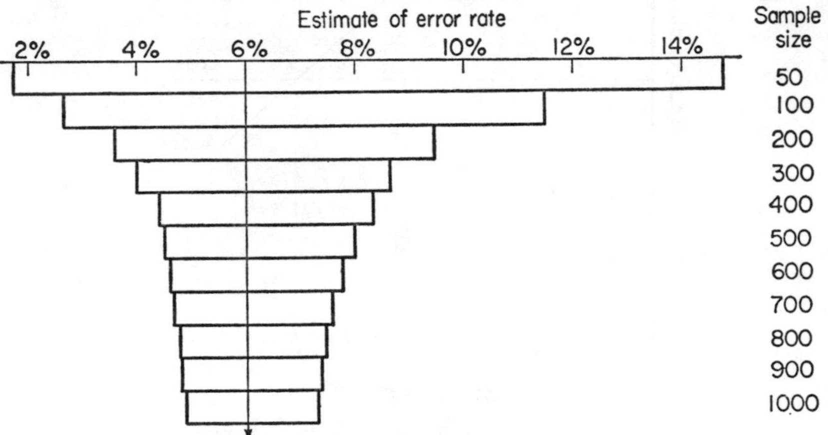


Fig. 4.2. Accuracy of actual error rate in population from various sample sizes.

| | |
|---|---|
| Size of population | 10,000 |
| Confidence level | 90% |
| Actual error rate | 6% |

A precise calculation of optimal sample size may be practically impossible but a rough calculation is sufficient to eliminate the type of non-optimal sample size described in the previous two paragraphs.

Figure 4.2 presents the optimal sample size problem in a slightly different format. The $x$ axis describes the precision limits on the estimate of the actual error rate, in this case 6%. Notice that as the sample size increases from 50 units (0·5% of population) to 1000 units (10% of population) the accuracy of inference increases *but at a diminishing rate.* Beyond a sample of 500 units the increase in accuracy is very small.

Each increase in sample size reduces the width of the precision limit at a fixed level of confidence. However, as each equal increment increases the sample size it provides a smaller *proportionate* improvement in the accuracy of the inference.

We can buy increased accuracy of inference by increasing the sample size but is the increased accuracy worth the cost? The answer to this

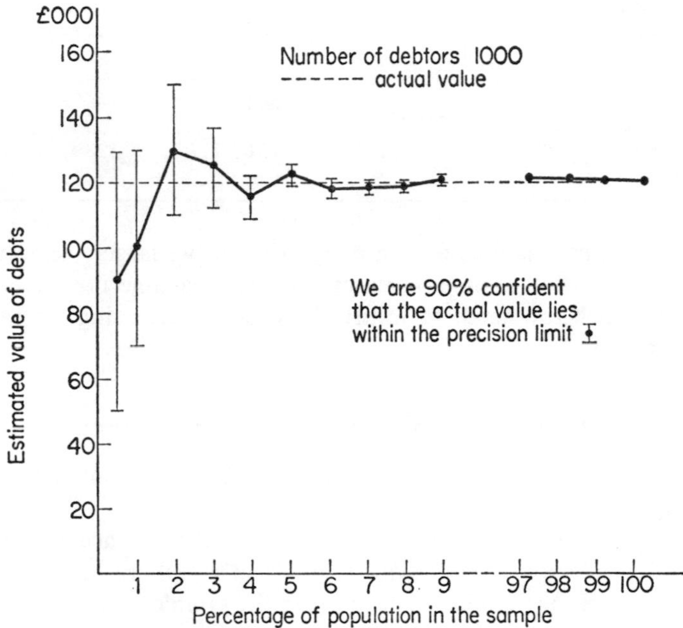


Fig. 4.3. Sampling to estimate the value of debtors. The estimate zeros in on the actual value of £120,000 as the sample size increases, but additional increments in the sample size provide steadily diminishing returns in improving the estimate.

question depends upon the auditor's judgement of the consequences of the inference being wrong.

Figure 4.3 provides a further exposition of the same point. The actual value of a population of debts is £120,000. As the sample increases in size the inference from the sample gets closer to the actual value but with each increment the accuracy increases at a diminishing rate.

In the previous examples we kept all of the factors constant except sample size and precision limit. We were thus able to trace the effect of an altered sample size on the precision limit. By keeping all factors constant except sample size and confidence level we can note the effect of an increment in sample size on the level of confidence.

This is shown in Table 4.1 below.

Table 4.1. An increasing sample size increases the level of confidence but at a diminishing rate. Population size 10,000, estimated error rate 2%, precision limit ±1%.

| Confidence level required | Percentage increase in initial confidence level | Sample size | Percentage increase in initial sample size |
|---|---|---|---|
| 90% | — | 504 | — |
| 95% | 5% | 700 | 39% |
| 99% | 9% | 1151 | 128% |
| 99·7% | 9·7% | 1474 | 192% |

In summary; as a sample from a population grows larger the accuracy of inference from the sample increases. But after a time the increase in accuracy is much too small to justify the cost of increasing the sample size further.

### THE EFFECT OF POPULATION SIZE ON SAMPLE SIZE

The purpose of statistical sampling is to make inferences about populations from samples drawn from these populations.

The key point in the economics of sampling is that the accuracy of the inference depends upon the *absolute* size of the sample rather than upon the proportion of the population included in the sample. This point is not, I suspect, intuitively obvious to most non-statisticians.

Suppose, for example, that we are trying to estimate the percentage of clerical workers employed by four firms. We decide to estimate by sampling. The number of people employed by the four firms is:

| Firm | Number employed |
|---|---|
| A | 10,000 |
| B | 50,000 |
| C | 100,000 |
| D | 500,000 |

We want to ensure that our estimate of the percentage of clerical employees is within 3% either side of the actual percentage with a 90% level of confidence. What size sample do we need to fulfil these requirements?

It might be thought that if the sample size that meets these requirements for firm A is, say, 800, then the sample sizes required for firms B, C and D will be 4000, 8000, and 40,000 respectively, i.e. that to maintain the same level of accuracy of inference we must take an equal *proportion* of each population. *This is not so.* The accuracy of the inference depends mainly on the *absolute* size of the sample, the population size plays a relatively minor role in determining the accuracy of the inference.

The actual sample sizes to give the required degree of confidence in the inference for estimating the percentage of clerical employees are:

| Firm | Population size | Percentage increase on initial population | Sample size | Percentage increase on initial sample size |
|---|---|---|---|---|
| A | 10,000 | — | 859 | — |
| B | 50,000 | 500% | 923 | 7% |
| C | 100,000 | 1000% | 931 | 8% |
| D | 500,000 | 5000% | 938 | 9% |

An increase in the population by a factor of 5000% increases the sample size by only 9%!

When a population is relatively small so that the sample becomes a significant proportion of the population, say in excess of 10%, then population size plays a more important role in determining the degree of confidence one can place in the inference. However, if we ignore this factor we are being *conservative* in our estimate of the accuracy of the inference. That is we are underestimating our confidence, or to put it another way we could achieve the confidence we require with a *smaller* sample size.

In my experience it is relatively rare for an auditor to check in excess of 10% of a population. We will discuss the point further in Chapter 11.

In summary; population size plays a very minor role in determining sample size. The accuracy of inference from a sample depends upon the absolute size of the sample and not upon the proportion of the population included in the sample.

*This conclusion is, perhaps, the most important contribution which statistics has made to the science of auditing.*

## WHAT LEVEL OF CONFIDENCE DOES AN AUDITOR NEED?

In Table 3.2 above we noted that if the precision limit, etc., are kept constant sample size is a function of the level of confidence. What level of confidence does an auditor need to have in an inference?

Traditionally statisticians work with levels of confidence of 90%, 95%, 99%, and 99·7%.

The level of confidence chosen depends firstly upon the consequence of the inference being wrong. If an invalid inference could have serious consequences the auditor may opt for a 99% level of confidence. If the consequence of the inference being wrong is not so serious, the auditor might opt for a 90% level of confidence.

The level of confidence chosen by the auditor also depends upon his assessment of the quality of the internal control system operated by the company being audited. If internal control is very good the auditor will require a relatively low level of confidence in his inference about the accuracy of accounting populations generated by this control system. In the U.S.A., where internal control is exceedingly good, the auditor has used levels of confidence as low as 60%![1] This, however, is rare and I doubt whether the average auditor will ever use confidence levels below 75% or 80%. Where internal control is non-existent the auditor may require a 95% or even 99% level of confidence.

In the early stages of using statistical sampling it might be advisable to use a level of confidence one step up from the one you will use when the method has been tried and tested.

But why, the reader might ask, should an auditor not use the 'safest' confidence level of 99·7% all of the time? The reason is illustrated in Table 4.1 above. A small increase in the level of confidence calls forth a much larger proportionate increase in sample size. A 5% increase in the level of confidence, for example, called forth a 39% increase in sample size.

Clearly the auditor will always work with the *minimum* level of confidence that is reasonable under the circumstances of the audit.

## THE PRECISION LIMIT

In Table 3.2 in Chapter 3 we constructed various confidence intervals 9 to 11, 8 to 12, etc., and we calculated the probability of the number of defectives falling within this range.

When the range is applied to an estimate we call it a *precision limit* or confidence interval.

For example we might wish to have a 90% level of confidence in the statement that the number of defectives in a population of 1000 units lies between (and including) 5% and 15%. This range, 5% to 15%, is our confidence interval for this problem. The precision limit is ±5%.

[1] See Chapter 17.

It is of the nature of auditing and accounting problems that the precision limit will always be very narrow, usually plus or minus 2% or less and seldom exceeding plus or minus 5%. Some auditors who use statistical sampling methods fix their precision limits for *all* jobs at ±2% except under special circumstances.

Table 4.2. Two examples of how sample size varies with confidence level and precision limit.

(a) Sample sizes at various levels of confidence and various precision limits—estimation sampling of attributes. Population 10,000, estimated error rate 3%.

| Precision limit | Sample size at $x$% level of confidence | | |
|---|---|---|---|
| | 99% | 95% | 90% |
| ±1% | 1617 | 1005 | 730 |
| 2% | 460 | 272 | 193 |
| 3% | 210 | 123 | 87 |
| 4% | 119 | 69 | 49 |
| 5% | 77 | 45 | 32 |

(b) Sample sizes at various levels of confidence and various precision limits—estimation sampling of variables. Population 10,000, standard deviation £2, estimated average value £10.

| Precision limit | Sample size at $x$% level of confidence | | |
|---|---|---|---|
| (plus or minus) | 99% | 95% | 90% |
| £1000 | 2103 | 1332 | 977 |
| £2000 | 624 | 370 | 263 |
| £3000 | 287 | 168 | 119 |
| £4000 | 164 | 96 | 68 |
| £5000 | 106 | 62 | 44 |

If the precision limit is fixed the risk factor can only be altered by manipulating the level of confidence. This might be considered an advantage since the auditor can now concentrate his attention on a single controllable variable.

## CONFIDENCE LEVEL VERSUS PRECISION LIMIT

If the precision limit is *not* fixed the confidence level and the precision limit can be traded off against one another. A higher confidence level can be obtained by widening the precision limit at a fixed sample size. Table

4.2 illustrates the sample sizes required in a given problem when we alter the confidence level and precision limit.

The reader will notice the considerable reduction in sample size as the precision limit grows wider. For example:

| Precision limit | Approximate percentage reduction on 1% limit |
|---|---|
| ±1% | — |
| 2% | 72% |
| 3% | 87% |
| 4% | 93% |
| 5% | 95% |

Note particularly the fall in sample size as the precision limit widens from 1% to 2%.

We can carry out a similar analysis on confidence level.

| Confidence level | Approximate percentage reduction on 99% level of confidence |
|---|---|
| 99% | — |
| 95% | 40% |
| 90% | 60% |

Notice that if we attempt to achieve an inference with above a 95% level of confidence or within a 2% precision limit there is a quite dramatic increase in sample size.

## SUMMARY

The auditor will normally use a level of confidence of between 80% and 95% and a precision limit of around ±2%. However, the final decision depends upon the particular circumstances of each case.

Population size plays a relatively small part in determining sample size.

## QUESTION SERIES 4

1. Sample size depends upon what four factors?
2. At what point does it become uneconomic to increase our sample size further?

3. Does the size of the population from which a sample is drawn have much effect on the accuracy of the inference from the sample? (*)
4. An estimate of a population proportion is being made under the following conditions.

| | |
|---|---|
| Population size | 5000 |
| Level of confidence | 95% |
| Precision limit | ±5% |
| Expected proportion | 20% |
| Required sample size | 234 |

Guess the sample size required if, other conditions remaining the same, the population size had been.

| | |
|---|---|
| 20,000 | ? |
| 100,000 | ? |
| 1,000,000 | ? |

(*)

5. What levels of confidence is an auditor likely to use?
6. Why does an auditor not always use a very high level of confidence?
7. What is the difference between a precision limit and a confidence interval? (*)
8. The auditor will normally use a precision limit of what range?
9. Sample size increases rather dramatically if the level of confidence is pushed above what percentage or if the chosen precision limit is less than what percentage?
10. 'Accuracy and reliability of inference work against one another in sampling.' True or false? (*)

## SOME ANSWERS TO QUESTION SERIES 4

3. Not much!
4. (1) 243 (2) 245 (3) 246
7. The terms mean much the same thing. A confidence interval must be symmetric around the mean to be called a precision limit.
10. True. The wider the confidence interval the higher the confidence level. The bigger the target the higher our confidence in hitting it.

# 5

# How to Measure the Standard Deviation of Accounting Populations

## INTRODUCTION

In the previous chapter we discussed several factors which determine the size of sample we need to make an inference about a population. In this chapter we will discuss another factor called the *standard deviation* of the population.

Standard deviation is one measure of a more general characteristic called the *dispersion* of the population. We will first examine the concept of dispersion and later describe several methods of calculating the standard deviation.

## THE CONCEPT OF DISPERSION

Table 5.1(a) provides a series of values from one inventory. Table 5.1(b) provides a series of values from another inventory.

The *frequency distribution* of each series is plotted in Figure 5.1(a) and 5.1(b). This frequency distribution shows us the number of times each value occurs. Let us call the value allotted to each item of inventory a *reading*.

We notice that Figures 5.1(a) and 5.1(b) have a number of interesting characteristics. First we notice that the mean value, that is the total value divided by the number of readings, of both distributions is identical at 9·92. Second we notice that both distributions are highly symmetrical, that is the number of readings falls away symmetrically on either side from the peak of the distribution.

Yet Figures 5.1(a) and 5.1(b) clearly show that the two distributions are not identical. Figure 5.1(a) has a higher peak than Figure 5.1(b) and Figure 5.1(b) has a much wider base than Figure 5.1(a). These two

Table 5.1 (a) and (b). The value of the items of inventory.

| Value of item of inventory £ | No. of items having this value (a) | (b) |
|---|---|---|
| 3 | — | 1 |
| 4 | — | 1 |
| 5 | — | 4 |
| 6 | 2 | 2 |
| 7 | 4 | 4 |
| 8 | 7 | 5 |
| 9 | 14 | 13 |
| 10 | 21 | 14 |
| 11 | 15 | 13 |
| 12 | 7 | 7 |
| 13 | 3 | 4 |
| 14 | 1 | 2 |
| 15 | — | 2 |
| 16 | — | 1 |
| 17 | — | 1 |
| | 74 | 74 |
| Mean value | £9·92 | £9·92 |

characteristics indicate that the readings represented in Figure 5.1(b) are *dispersed* more widely around the mean value of the population than the readings represented by Figure 5.1(a).

We need to measure this factor of dispersion of the readings around the mean value of a population when estimating the value of a population variable from a sample drawn from that population.

As a small experiment the reader may like to select several random samples of four units from the Table 5.1(a) and 5.1(b). Take the mean value of these samples by dividing their total value by four, and compare this sample estimate with the actual population mean of 9·92. The reader should find that if he takes say, a dozen samples of four units, then more of the estimates drawn from Table 5.1(a) will be closer to the actual population mean value than those drawn from Table 5.1(b).

I selected a dozen random samples of four units from each population and the results were as follows:

| Mean estimate from table | 1 | 2 | 3 | 4 | 5 | 6 | 7 | 8 | 9 | 10 | 11 | 12 |
|---|---|---|---|---|---|---|---|---|---|---|---|---|
| 5.1(a) | 10 | 10 | 11·75 | 11 | 10·75 | 11 | 9 | 9 | 9 | 10·25 | 8.75 | 9 |
| 5.1(b) | 10·5 | 9·75 | 12·75 | 11·5 | 11·50 | 9·75 | 8·5 | 8·5 | 8·5 | 10·5 | 8·5 | 8·5 |

From the 24 estimates made, those drawn from Table 5.1(a) provided the closest estimates.

This is because Table 5.1(a) has a narrower dispersion than Table 5.1(b), therefore more of the readings in Table 5.1(a) are closer to the population

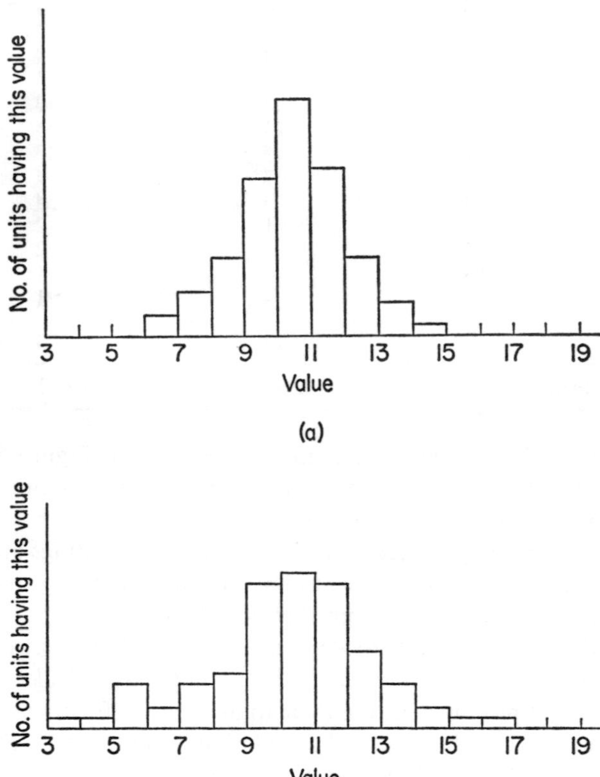


Fig. 5.1. Frequency distributions of two series of inventory values. Notice that inventory distribution (a) has a higher peak and a narrower base than (b), i.e. it has a narrower dispersion.

mean. Therefore for a given sample size the population with the narrower dispersion will give a closer estimate of the population mean.

Since the dispersion of the population affects the sample size required the auditor needs to measure, or at least to estimate, the dispersion of the population he is auditing when he is estimating a *value* such as inventory value, debtors value or error value. The measure of dispersion that the auditor will use is called the *standard deviation* of the population.

We will now describe three methods of calculating the standard deviation. The first two methods provide a precise measure of the standard deviation of the population, and can be used when the population to be audited is relatively small. When, as in most auditing applications, the population to be audited is relatively large, consisting of several thousand units or more, then the auditor will use the third method described which *estimates* the standard deviation.

This third method, called the *average range method*, will almost always be used by auditors since most accounting populations are very large.

## THE CONVENTIONAL METHOD OF CALCULATING THE STANDARD DEVIATION OF A POPULATION

Table 5.2 presents ten readings of a variable, say the height of ten males. The conventional method of calculating the standard deviation of this population is as follows:

1. Calculate the mean of the population of $n$ units.
2. Subtract the mean from each of the readings.
3. Square each of the results arrived at under (2).
4. Add these squared numbers together and divide the result by $n$.
5. Find the square root of the answer arrived at under (4).

The answer arrived at under (5) is the standard deviation of the population.

Table 5.2. The height of ten males.

| Male | Height (in inches) |
|---|---|
| 1 | 68 |
| 2 | 70 |
| 3 | 65 |
| 4 | 71 |
| 5 | 67 |
| 6 | 68 |
| 7 | 66 |
| 8 | 68 |
| 9 | 72 |
| 10 | 64 |

For the data presented in Table 5.2 the calculations are as follows:

1–3

| (a) Height | (b) Mean | (c) (a)–(b) | (d) (c)$^2$ |
|---|---|---|---|
| 68 | 68 | 0 | 0 |
| 70 | 68 | +2 | 4 |
| 65 | 68 | −3 | 9 |
| 71 | 68 | +3 | 9 |
| 67 | 68 | −1 | 1 |
| 68 | 68 | 0 | 0 |
| 66 | 68 | −2 | 4 |
| 68 | 68 | 0 | 0 |
| 72 | 68 | +4 | 16 |
| 64 | 68 | −4 | 16 |
| | | | 59 |

4. $59 \div 10 = 5{\cdot}9$
5. $(5{\cdot}9)^{1/2} = 2{\cdot}43$

The standard deviation of the population of the height of ten males is 2·43 inches.

## A SPEEDIER WAY OF CALCULATING STANDARD DEVIATION

The method of calculating standard deviation described above is rather long-winded if the population is divided into a large number of classes.

If a desk calculator is available the following method is a good deal simpler.

1. Square each value in the population, add the squared values together and divide by $n$, the number of units in the population. Call the result of this calculation $a$.
2. Sum each value in the population, divide the result by $n$, then square the answer. Call the result of this calculation $b$.
3. Subtract $b$ from $a$ and find the square root of the answer.

The end result of (3) is the standard deviation of the population.

For the data presented in Table 5.2 the calculations are as follows:

1.

| Value | $(\text{Value})^2$ |
|---|---|
| 68 | 4624 |
| 70 | 4900 |
| 65 | 4225 |
| 71 | 5041 |
| 67 | 4489 |
| 68 | 4624 |
| 66 | 4356 |
| 68 | 4624 |
| 72 | 5184 |
| 64 | 4096 |
| 679 | 46,163 |

2. $46{,}163/10 = 4616{\cdot}3$
   $679/10 = 67{\cdot}9,\ (67{\cdot}9)^2 = 4610{\cdot}4$
3. $(4616{\cdot}3 - 4610{\cdot}4)^{1/2} = (5{\cdot}9)^{1/2} = \pm 2{\cdot}43$

We ignore the negative root and find 2·43 to be the standard deviation of the population, the same answer we arrived at by using the conventional method.

The reader may think the second method even more complicated than the first. If, however, the reader has access to a desk calculator he will find that the second method is easier to handle.

The standard deviations of the two populations set out in Table 5.1(a) and (b) turn out to be:

(a) 1·63
(b) 2·67

The reader may like to check this using the two methods set out above. Remember that once you have subtracted the mean from each of the readings and squared the result you must multiply each of the results by the number of items in the class interval.

## FINDING THE STANDARD DEVIATION OF LARGE POPULATIONS

The two methods of calculating the standard deviation described above gives a precise measure of the standard deviation. These methods are not, unfortunately, of much use in calculating the standard deviation of most accounting populations.

Most accounting populations consist of tens of thousands of units and so, unless we can classify the population on a computer, the previous methods of calculating the SD are much too laborious.

We cannot, then, make a *precise* calculation of the SD; the best we can do is to *estimate* the standard deviation.

The auditor can draw a sample, calculate the standard deviation of this sample, then use this statistic as an *estimate* of the standard deviation of the population from which it is drawn. Table 3.4 on page 23 describes the formula for calculating the standard error of the estimate of standard deviation. The *average range method*, which we shall now describe, is another way of estimating the standard deviation of a very large population. The method was first suggested by F. E. Grubbs and C. L. Weaver in 1947.

## AVERAGE RANGE METHOD OF ESTIMATING STANDARD DEVIATION

The procedure for estimating the standard deviation of a population using the average range method is as follows:

1. Select a sample of 49 units at random from the population.
2. Write the 49 values down *in the order in which they are drawn.* Do *not* sort them into sequence.
3. Segregate the 49 units into 7 groups of 7.
4. Calculate the *range* of each group by subtracting the smallest value from the greatest value in that group.
5. Calculate the average range by dividing the sum of the 7 ranges by 7.
6. Divide the average range by the number 2·704.
7. The answer is a good *initial* estimate of the standard deviation. Later we will test the accuracy of this estimate using the total sample.

The various instructions are clear except for instruction 6. Where does the mystical number 2·704 come from? Suppose that we had two numbers near the extreme upper and lower bounds of the Normal distribution. We learn that 99·7% of the area under the Normal distribution lies three standard deviations either side of the mean. Thus we learn that these two numbers must be roughly six standard deviations apart. Therefore if we subtract the smallest from the largest and divide this range by 6 we have a rough estimate of the standard deviation!

The same principles can be applied to estimating the standard deviation from narrower ranges. Using groups of 7 we can use statistical methods to calculate the *most likely* distance between the upper and lower bound of this average range in standard deviations. In this case the estimate is that the upper and lower bounds are 2·704 standard deviations apart.[1]

Let us now work through an actual example.

1. A population consists of 638 units. We select 49 random numbers between 001 and 638 and pull out the following random values.

| (1) | (2) | (3) | (4) | (5) | (6) | (7) |
|---|---|---|---|---|---|---|
| 536 | 411 | 276 | 574 | 274 | 684 | 717 |
| 481 | 388 | 385 | 386 | 318 | 721 | 608 |
| 211 | 887 | 449 | 255 | 914 | 702 | 212 |
| 917 | 202 | 350 | 319 | 382 | 684 | 847 |
| 354 | 318 | 617 | 417 | 617 | 191 | 363 |
| 626 | 514 | 848 | 891 | 493 | 227 | 666 |
| 255 | 800 | 226 | 902 | 550 | 543 | 510 |

2. The range of each group is:

| | (1) | (2) | (3) | (4) | (5) | (6) | (7) |
|---|---|---|---|---|---|---|---|
| Highest | 917 | 887 | 848 | 902 | 914 | 721 | 847 |
| Lowest | 211 | 202 | 226 | 255 | 274 | 191 | 212 |
| | 706 | 685 | 622 | 647 | 640 | 530 | 635 |

3. The average range is therefore:

$$\frac{4465}{7} = 637·9$$

4. The estimate of standard deviation is therefore:

$$\frac{637·9}{2·704} = 235·9$$

Note, however that this is only an *initial* estimate using 7 groups of seven random units. The more groups of seven we take the more precise is the estimate of standard deviation likely to be. If our required sample size turned out to be 154 units, this would give us 22 groups of seven units to estimate the SD. But remember that the 154 units must be listed in the order in which they appear in the random number table.

### THE EFFECT OF STANDARD DEVIATION ON SAMPLE SIZE

Earlier in this chapter we noted that other things being equal an increase in standard deviation will increase the required sample size, and vice

[1] See Grubbs and Weaver (1947).

versa. Suppose for example that the following sampling problem presents itself.

| | |
|---|---|
| Population size | 10,000 units |
| Level of confidence required | 95% |
| Precision limit | ± £10,000 |
| Unit precision limit | ± £1 |

Table 5.3 presents the sample size required to satisfy the above requirements for various standard deviations.

Table 5.3. The table illustrates the effect of standard deviation on sample size.

| Standard deviation | Percentage of initial SD | Sample size required | Percentage of initial sample size |
|---|---|---|---|
| £50 | — | 4900 | — |
| £25 | 50 | 1936 | 40 |
| £10 | 20 | 370 | 8 |
| £5 | 10 | 96 | 2 |
| £3·3 | 7 | 43 | 1 |

A reduction in the size of the standard deviation causes an even greater proportionate reduction in sample size.

Now we recall that in the case of the confidence level and the precision limit the auditor could reduce his sample size by moving to a *lower* confidence level or a *wider* precision limit, but it may seem that he cannot do this in the case of standard deviation. *The standard deviation of the population is given, it cannot be altered.*

This latter statement is true, but fortunately we can get around the problem by altering the population itself. We can *stratify* the population into two or more subpopulations in such a way that the standard deviation of the subpopulation of small value is much lower than the SD of the whole population. This allows us to reduce sample size.

We shall return to the problem of stratification in a later chapter.

## THE COEFFICIENT OF VARIATION

A simple measure of the relative dispersion of a population of readings around the mean value is called the *coefficient of variation.*

The C of V is simply the standard deviation divided by the mean. For example if the SD is £100 and the mean is £300 the C of V is 0·33.

## SUMMARY

The dispersion of the accounting population around the mean value is another important determinant of sample size. The measure of dispersion used by the auditor is called the standard deviation of the population.

Various methods of calculating a precise value for the standard deviation are available. However most accounting populations are very large and the auditor will normally use the average range method to *estimate* the standard deviation of the population.

An increase in standard deviation will call forth a more than proportionate increase in sample size, and vice versa. We can control the standard deviation of the population to some extent by stratifying the population into several subpopulations.

## QUESTION SERIES 5

1. What do we mean by the term *dispersion* of a population?
2. From the data provided in Table 5.1 on p. 37 select 5, 10, 15 and 20 units from column (a) and column (b). Calculate the mean value of each of these samples. Do the larger samples provide a closer estimate of the population mcan? (*)
3. Why does a population with a smaller dispersion require a smaller sample size other things being equal?
4. What is the name of the most commonly used measure of dispersion?
5. Use the two conventional methods of calculating standard deviation to calculate the SD of thc following population

   12, 9, 7, 15, 8, 4, 9, 2, 21, 15. (*)

   Compare the results.
6. Why are the two methods of calculating SD used in question 5 above of little value for calculating the SD of most accounting populations?
7. Use the average range method to estimate the SD of the 100 values in Table 7.2 on p. 74 (*)
8. How does standard deviation affect sample size?
9.

| | |
|---|---|
| Population size | 5000 units |
| Level of confidence | 90% |
| Precision limit | ± £50,000 |
| Unit precision limit | £10 |
| Estimated mean value of unit | £100 |

| Standard deviation £ | Required sample size |
|---|---|
| 300 | 1880 |
| 200 | ? |
| 100 | ? |
| 50 | ? |

Guess the sample size required for SDs of £200, £100 and £50. (*)

10. An auditor discovers that the standard deviation of the population he is auditing is very large relative to the mean. Suggest a method of coping with this problem. (*)

## SOME ANSWERS TO QUESTION SERIES 5

2. On the average they will.
5. 5.38
7. 23 (approx.)
9. (1) 890 (2) 257 (3) 69
10. Stratify the population.

# 6

# Methods of Drawing a Random Sample

## INTRODUCTION

The topic of selecting a viable random sample from an accounting population deserves a chapter to itself.

This topic is *not* the most difficult part of applying statistical sampling to auditing, but it seems to throw up the most problems to the inexperienced practitioner.

## WHAT IS A SAMPLE?

A *sample* is any subset of a population. At one extreme a sample might consist of a single unit from a population. 'Depth auditing' of a single item through various accounting procedures is an example of a unit sample. At the other extreme a sample could consist of the entire population! However, in practice, the purpose of sampling is to make an inference about some characteristic of a population without having to examine the entire population.

In auditing and control applications it is rare for the sample size to exceed 10% of the population, and it is usually much less than this.

## DRAWING THE SAMPLE

There are two basic methods of drawing a sample from a population. The first of these is called *judgement* sampling, the second *random* sampling.

To draw a judgement sample from a population the sampler must decide *in advance* on the criteria he will use for selecting the sample. If, for example, he is trying to detect invalid authorizations of expenditure he may decide to select out all of the items of expenditure authorized by Mr. X because he suspects, from information received elsewhere, that Mr. X is careless in signing authorizations. Alternatively an auditor, while auditing the accuracy of a company's payroll, might decide to audit the

July pay slips because he knows that temporary staff were employed in the pay office during that period.

In both of the above examples every unit in the population did not have an equal chance of selection on each draw. The sampler decided in advance on the criteria for accepting or rejecting a unit into the sample.

The key factor in *random* sampling is that on each draw every unit remaining in the population has an *equal chance of selection.* If the population consists of 10,000 units then on each draw, with replacement, every unit has an equal chance of selection.

*The apparatus of statistical sampling can be used to make inferences about populations from samples, if, and only if, the sample is a random sample.*

## RANDOM SAMPLING WITH AND WITHOUT REPLACEMENT

Random sampling can be carried out with or without *replacement.* If an auditor is checking 10,000 invoices and he draws one invoice using random sample methods the probability that any single invoice will be selected is 1 chance in 10,000. If that invoice is checked and then laid aside the probability that any single invoice will be selected on the next random draw is 1 chance in 9,999 and so on. This method of sampling is called random sampling *without* replacement. If the unit selected is placed back into the population so that it can be selected again, the probability of selecting any one invoice will always be one in 10,000. This second procedure is called sampling *with* replacement. Most sampling procedures in auditing and accounting use random sampling *without* replacement. If the sample size is small relative to the size of the population, which is usually the case in auditing and accounting work, the difference between the *with* or *without* replacement methods is not important. But when the sample size exceeds, say, 10% of the population the sampler should take care, since most sampling tables are calculated on the basis of random sampling *with* replacement.

## JUDGEMENT VERSUS RANDOM SAMPLING

Some accountants have compared the job of the auditor to the job of the geologist seeking for oil. The geologist, so these accountants argue, does not take *random* borings to find oil, he uses his experience and technical expertise to drill in those places where he expects to find oil. Similarly, it is argued, the auditor would be foolish to use random sampling methods to seek inaccuracy or fraud when his experience and technical expertise can lead him to those areas where he is most likely to find fraud.

This argument shows a misunderstanding of the scientific sampling method. If one part of a population is thought to be different from another part so far as the given condition—inaccuracy and fraud—is concerned, then the population is not *homogeneous.* As we explained in Chapter 1, if we have prior information which leads us to believe that, say, the error rate in one branch or type of document, etc., is greater or less than elsewhere we must *stratify the population* accordingly so that each *stratum* is homogeneous. We will then adopt a different sampling plan for each strata. Our confidence level, and therefore our sample size, is likely to be different for each strata.

Random sampling, then, is applied to a homogeneous population. The auditor's experience and technical expertise are used to stratify a population into homogeneous strata, each strataum becoming a separate population for sampling purposes.

## THE ADVANTAGES OF RANDOM SAMPLING

As stated above the sophisticated apparatus of statistical inference can only be used when the inferences are based on a random sample. Random sampling, however, enjoys other advantages over judgement sampling.

Since no one, not even the auditor himself, can predict the sample he will choose in advance, a potential misfeant cannot select an area of the audit population with a low probability of being audited. In these days of highly programmed audit procedures this is a useful benefit. Smurthwaite (1965) for example, points out in a study of his firm's auditing procedures that the audit staff tended to select a month for auditing in year $n$ which was either side of the month selected in the previous year $(n - 1)$.

Random sampling also overcomes any *unconscious* tendency on the part of the auditor to select units of the population which have certain characteristics or to ignore other units with differing characteristics. For example some auditors tend to concentrate on hand-written as against typed documents when selecting units to be audited.

## THE PURE RANDOM SAMPLE

The most efficient form of sample, in the sense that it provides the best prediction about the population, is a pure random sample.

A sampler could use a digital computer or roulette wheel or some other type of random number generator to provide his own random numbers, but the simplest and most effective procedure is to use a table of random numbers prepared and tested by a competent authority.

Many tables of random numbers are now available for use by the sampler. A series of random numbers is provided in this book on pp. 266–269. Table 6.1 provides a section of these tables.

Table 6.1. Section from table of random numbers.

| | | | |
|---|---|---|---|
| 87421 | 69113 | 28417 | 66841 |
| 39265 | 98992 | 37841 | 61624 |
| 36421 | 16405 | 95106 | 19263 |
| 84033 | 36705 | 41606 | 72103 |
| 68280 | 87750 | 19742 | 76315 |

The procedure for using random number tables is as follows.

Suppose we wish to select 82 random numbers between 837 and 2154. How do we go about it?

First we transform 837 into 0837. Thus all the numbers we select will have four digits.

Second we start *at any point* in the random number table—we can use random numbers to select the starting point—and move *systematically* in any direction collecting a series of four digit numbers. Suppose we start at the left-hand end of row three on Table 6.1 and move to the right. The random numbers come out to be 3642, 1164, 0595, 1061, 9263, 8403, 3367, 0541, 6067, 2103, 6828, 0877, 5019, 7427, . . . .

Now we wish to select out 82 random numbers between 0837 and 2154. Only four of the numbers so far chosen 1164, 1061, 2103, and 0877 fall into this category. We must continue selecting random numbers until we find 82 numbers lying between 0837 and 2154. Duplicate numbers will be rejected if we are sampling without replacement.

Once we have found 82 random numbers lying between, and including, 0837 and 2154 we sort them into sequence and we are now ready to select our random sample.

If we assume that the population we are auditing is a set of inventory bin cards numbered 837 to 2154, we now select out those bin cards corresponding to the 82 random numbers selected. We can now use this random sample of bin cards to make inferences about the population of 1318 bin cards.

## THE SYSTEMATIC RANDOM SAMPLE

An alternative to pure random sampling is systematic random sampling. Under certain circumstances this latter method may enjoy an *operational* advantage over pure random sampling.

Systematic random sampling can best be illustrated by an example.

If we return to our previous example of selecting 82 bin cards at random between 0837 and 2154, using systematic random sampling methods we would divide 318 (the number of units in the population) by 82 giving 3·88 to the nearest decimal place. This number is close to 4. Let us then take every *fourth* item in the population starting at a random point in the first group of four. This will give us 79 random units and since we need 82 we will either pick another three at random from the population or alternatively we can divide the population into roughly three sections and select one unit at random from each section.

In formal language. If the population consists of $n$ units and we want a random sample of $s$ units. We select every $n/s = r$th unit systematically starting at a random point among the first $r$ units in the population.

If a population is not listed in numerical sequence it is much easier to select a random sample using systematic rather than pure random sampling.

The danger in using systematic random sampling is that every $r$th unit *might* correspond to an existing sequence in the population so that the sample items are always being drawn from the same section of a recurring pattern.

Suppose for example that a payroll of 80 employees is listed as follows

| | |
|---|---|
| 01–05 | high-salaried employees |
| 06–20 | middle-salaried employees |
| 21–80 | low-salaried employees |

and these employees are paid weekly so that the population of pay slips is $50 \times 80 = 4000$ pay slips. If $r$ is close to 80, which is not improbable in this case, the sampler might select 50 units from the same strata of the population giving a highly misleading sample.

With most accounting populations the coincidence is unlikely but when systematic sampling is used the sampler should always examine the data to exclude this possibility.

The MUS system described in Chapter 17 makes an interesting and original use of systematic sampling.

## JUNIOR DIGIT RANDOM SAMPLING

Another form of random sampling can be used when each unit in the population is stamped or otherwise associated with a number. Invoices, for example, are often stamped in number sequence when they are processed.

Suppose, for example that an auditor wishes to select a sample of 100 from a population of 1000 invoices stamped 0001 to 1000, *or any other number pattern, not necessarily sorted in sequence.*

By the simple process of selecting every invoice with a given number, say 8, in the junior, i.e. right-hand end, of the invoice number he will select approximately 10% of the population, i.e. approximately 100 units.

If he wishes to select 50 random invoices he could use the formula, 'select every invoice with 8 as the junior digit but only if it is preceded by an even number.' This will provide a sample of approximately 50 units.

A random sample of any other percentage of the population can be drawn by using a little ingenuity. For example a random sample of 70 from a population of 583 could be drawn by using the formula, 'select every invoice with an 8 in the junior digit position, also select every invoice with a 9 in that position but only if there is also a 1 or 2 in the tens position.' This will give us approximately $10\% + 10\%/5 = 12\%, 12\% \times 583 = 70$.

Junior digit sampling will not give us exactly the sample size we require, but only a close approximation. We can use some other random sampling method to top up the sample to the required size or to eliminate excess units.

Notice particularly that the numbers used need not be in any sequence.

Beware, however, against numbers which incorporate some kind of coded information, i.e. 8 in the junior digit position means that this customer is situated in Bradford. These are not suitable for junior digit random sampling.

## OTHER METHODS OF RANDOM SAMPLING

A wide variety of other methods are available for effecting a reasonable random sample.

In one case a sample of about 1000 documents was required from a population of 500,000 documents in the vault of an insurance company. The documents were bound into batches of unequal size. A pure or even systematic random sample based on policy number would have been tedious.

Instead the auditor selected his sample by measuring along the side of the batches. Suppose the batches were 10,000 inches long if placed end to end. Then the auditor selected a document every $10{,}000/1000 = 10$ inches along each batch, carrying any surplus inches to the next batch.

In another example a random sample of invoices stored on microfilm was effected by spinning the arm of the microfilm unit and stopping it about every five seconds. This method should provide a pretty good random sample.

The reader can probably think up other methods based on the same principle.

## TESTING FOR RANDOMNESS

Earlier in this chapter we introduced the idea of a random number table, a sequence of numbers which are random in the sense that no pattern can be detected in them. If we are given one number we cannot predict the next. The probability that the next number will be 0, 1, 2, . . . , 9 is always 1 chance in 10.

How do we test for randomness? In theory this is an insoluble problem. Numbers can be related in an infinite number of ways and we cannot test for the non-existence of every one of these patterns. However various tests have been devised by statisticians which give us a reasonable assurance that a series of numbers are in a relatively random sequence.

It is advisable for an auditor or accountant who is using statistical sampling methods to be acquainted with one of these methods.

## THE 'RUN' METHOD OF TESTING FOR RANDOMNESS

This method operates as follows.

1. Find the middle, or *median*, number in the group of numbers, i.e. with numbers 3, 6, 8, 9, 9, 11, 15, 17, the number 9·5 is the median or middle number.
2. Underline all of those numbers which are greater than the median number.
3. Count the number of runs, i.e. sequences of numbers which are continuously above (underlined) or below (not underlined) the median number.
4. Let

   $U$ = number of runs.
   $n$ = number of units underlined.

   Then by using Table 6.2 on page 54 we can be 95% confident that if $U$ falls between but not including the numbers cited for any given $n$ then the series of numbers is relatively random.

Table 6.2. Testing for randomness. If $U$ is between the numbers given for a given $n$, then there is a 95% chance that the series of numbers under investigation is random.

| $n$ | Is $U$ between? | |
|---|---|---|
| 5 | 2 | 10 |
| 6 | 3 | 11 |
| 7 | 3 | 13 |
| 8 | 4 | 14 |
| 9 | 5 | 15 |
| 10 | 6 | 16 |
| 11 | 7 | 17 |
| 12 | 7 | 19 |
| 13 | 8 | 20 |
| 14 | 9 | 21 |
| 15 | 10 | 22 |
| 16 | 11 | 23 |
| 17 | 11 | 25 |
| 18 | 12 | 26 |
| 19 | 13 | 27 |
| 20 | 14 | 28 |

*Example*

The following represent the value of sales of a product during fifty weeks. Is the value of sales in any one week a random process or is there, perhaps, a seasonal factor?

Table 6.3. Is this series of numbers random?

| | | | | | | | | | |
|---|---|---|---|---|---|---|---|---|---|
| 26 | 30 | 29 | 33 | 25 | 30 | 32 | 35 | 34 | 32 |
| $\underline{38}$ | 33 | $\underline{40}$ | $\underline{42}$ | 36 | $\underline{44}$ | $\underline{47}$ | $\underline{49}$ | $\underline{51}$ | $\underline{46}$ |
| $\underline{52}$ | $\underline{56}$ | $\underline{54}$ | $\underline{51}$ | $\underline{46}$ | $\underline{42}$ | $\underline{40}$ | $\underline{44}$ | $\underline{47}$ | $\underline{41}$ |
| $\underline{38}$ | 35 | 33 | 31 | 35 | 37 | $\underline{40}$ | 33 | 29 | 31 |

*Solution*

1. The median number is 37·5.
2. We underline all numbers greater than 37·5.
3. There are nine runs altogether
   Therefore
4. $U = 9$
   $n = 20$

Table 6.2 tells us that when $n = 20$ if $U$ is greater than 14 or less than 28 there is a 95% chance that the series is random. Since $U$, in this case, is 9 we conclude that the series is *not* random.

Let us look at another number series presented in Table 6.4.

Table 6.4. Is this series of numbers random?

| 41 | 65 | 96 | 17 | 34 | 88 | 20 | 28 | 83 | 53 |
|---|---|---|---|---|---|---|---|---|---|
| 83 | 40 | 60 | 59 | 36 | 29 | 59 | 38 | 99 | 82 |
| 95 | 78 | 29 | 34 | 78 | 17 | 26 | 77 | 09 | 43 |
| 20 | 85 | 77 | 31 | 56 | 70 | 28 | 42 | 43 | 24 |

*Solution*

1. Median is 43.
2. There are 21 runs.
3. $U = 21$
   $n = 19$
4. The mean and standard deviation are the same as in the previous problem. Therefore the limits are 27/13. Since 21 is less than 27 and greater than 13 we accept the hypothesis that this is random series of numbers—which in fact it is since the numbers are taken from a random number table.

## TESTING LARGER POPULATIONS

When $n$ exceeds 20, which is usually the case with auditing and accounting populations, we can treat $U$ as forming a Normal distribution, with

$$\text{mean} = (n + 1)$$

and

$$\text{standard deviation} = \left(\frac{2n(n-1)}{2n-1}\right)^{1/2}$$

and use a table of the Normal distribution, such as that on p. 238, for calculating the probability that the series is random.

When $n$ is very large the median must be estimated.

For example using the data in Table 6.3

$$\text{mean} = (n + 1) = 21$$

$$\text{standard deviation} = \left(\frac{2n(n-1)}{2n-1}\right)^{1/2}$$

$$s = \left(\frac{40(19)}{40-1}\right)^{1/2}$$

$$s = (19{\cdot}49)^{1/2}$$

$$s = 4{\cdot}415$$

Since 95% of the readings lie within 1·96 standard deviations of the mean the limits are

$$21 + (1{\cdot}96 \times 4{\cdot}415) = 29{\cdot}65$$

and

$$21 - (1{\cdot}96 \times 4{\cdot}415) = 12{\cdot}35$$

This compares with the more accurate, and narrower, limits of 28/14 in Table 6.1. The method using the Normal distribution will, however, become more exact as $n$ increases.

## SUMMARY

There are two basic strategies available for drawing a sample from a population. We can use either a judgement or a random sample.

Traditionally auditors use judgement sampling but the powerful methods of statistical analysis can only be used on a sample if it is a random sample.

The key point in random sampling is that each unit of the population must have an equal chance of selection on each draw.

Random sampling can be effected by (a) using numbers drawn from random tables, (b) systematic sampling of the population, (c) junior digit sampling, or (d) other methods depending on the auditors ingenuity.

A method of testing a series of numbers for randomness is explained.

## QUESTION SERIES 6

1. If the auditor decides to test all of the payslips for the first week in June is this a judgement or a random sample? (*)
2. What is the key characteristic of a random sample?
3. 'The total of all samples drawn from a population can never exceed the size of the population.' True or false? (*)
4. If we are testing a debtor's list for bad debts and have reason to believe that some dealers accept lower quality business than others should we take a *random* sample to test the bad debts provision? (*)
5. Suggest two other advantages of random sampling over judgement sampling.
6. Select twenty random numbers between 63 and 21714 from the random number tables on p. 268.
7. Using the data from question 6 calculate $r$ for a systematic random sample of 20 units. How do you select the first unit?
8. What is the main advantage and disadvantage of systematic random sampling?

9. Devise a formula using junior digit random sampling to select 1650 units from a population of 32,000. (*)
10. Suggest two other methods of drawing a random sample.
11. Test Table 6.1 on p. 50 to see if the series of 100 numbers is really random. (*)

## SOME ANSWERS TO QUESTION SERIES 6

1. It is a cluster sample. Whether it is a random or judgement sample depends upon whether the first week in June was chosen by random methods.
3. False, if sampling *with replacement* is used.
4. We should take a stratified random sample. The units within each strata being, so far as we know, of equal quality.
9. Select all units with a 1 in junior digit position except when the tens column contains an even digit. (This provides a sample of about 1600.)
11. They are relatively random.

# 7

# Some Comments on Defining and Analysing an Accounting Population

## WHAT IS A POPULATION?

A *population* or *field* is any group of items sharing a common characteristic. Examples of accounting populations are:

All of the debts owed to a company at a given point in time.
The various types of inventory stocked by a company at a given point in time.
All of the transactions recorded by a clerk during a given period of time.
The charges for all of the service jobs performed by a maintenance department during a given period of time.

Notice that accounting populations are made up of *either* a number of units existing at a given point in time, i.e. debts, *or* a number of units occurring between two points in time, such as a series of clerical operations. This distinction may be significant when selecting a sampling plan.

We noted above that the factor which binds a population together is the common characteristic shared by all of the units in the population. This characteristic of a population may be an attribute or a variable. Before we say more about defining and analysing populations let us be quite clear what we mean by the word *attribute* and *variable* as used in a statistical context.

## ATTRIBUTES AND VARIABLES

An *attribute* is a characteristic which a unit of a population either possesses or does not possess. It represents a quality rather than a quantity.

An accountant is either a chartered accountant or he is not. A debt owing is either overdue or it is not overdue. A payment is either properly authorized or it is not properly authorized.

In each of these examples every unit, i.e. the population of accountants, debtors or payments has, or does not have, a given attribute. The populations concerned are *binomial* populations, they can be split into two *classes*. One class, $p$, consisting of all of those units which possess the attribute, and one class, $q = 1 - p$, consisting of all of those units which do not possess the attribute.

The statistical technique which estimates the proportion of a population which possesses a given attribute is called *estimation sampling* of *attributes*.

Note that in estimation sampling of attributes each unit in the population must be allocated to *either* one class *or* the other. In other words the two classes must be *mutually exclusive*.

A *variable* is a property of a unit of a population which is measurable.

The value of an item of inventory, the number of days a debt is overdue, the cost of a service job, the value of a clerical error are all examples of variables.

The statistical technique which estimates the total value of a given variable of a population is called *estimation sampling of variables*.

An auditor checking an inventory list for errors might first wish to calculate the *proportion* of errors. To do this he would use estimation sampling of attributes. Secondly he might wish to estimate the *total value* of error. To do this he would use estimation sampling of variables.

Estimation sampling of attributes and variables are the two basic sampling techniques used by auditors when attempting to verify accounting populations.

## ON DEFINING ACCOUNTING POPULATIONS

With many sampling problems it is not easy to define what statisticians call the *sampling frame*. That is the population from which the sample will be drawn.

Fortunately, for our purpose, accounting populations are easy to define. Accountants are usually scrupulous in their classification procedures, and the various ledgers which make up most accounting populations are clearly delineated and not physically scattered over a wide area or mixed up with other populations.

However auditors should beware of compounding several populations together. If, for example, an auditor is verifying five different attributes of a population of invoices then each attribute makes up a separate population. It may be that if the accuracy of each attribute is independent of the others then the auditor can use the same sample to check all five attributes, but he must never forget that he is checking five distinct populations and not one.

We will return to this point later.

Once the population is defined and the frame of reference settled the auditor will be well advised to calculate certain parameters of the population.

We have already described how he can estimate two of these parameters, the mean and the standard deviation. Two other factors which the auditor ought to take into account are the *homogeneity* and the *skewness* of the population.

## THE MEANING OF HOMOGENEITY

If we decide to make an *inference* about some characteristic of a population by drawing a sample from that population, then, before we draw the sample, we must try to ensure that the population is relatively *homogeneous*. By homogeneous we mean that, *so far as we know*, any one part of the population is similar to any other part. Suppose, for example, that ten branches of a given retail store record approximately 10,000 sales transactions each month. We wish to estimate the error rate in filling in these sales slips. If we have no reason to suspect that the error rate in one branch differs significantly from the error rate in any other branch, we can treat the 100,000 sales slips as a single homogeneous population, and we draw our sample from these 100,000 sales slips. If, on the other hand, we suspect, that, say, one branch is run by an inefficient manager and may have a much higher error rate than the others, then the 100,000 population is *not* homogeneous, and we must segregate out the 'doubtful' branch for separate treatment. Baysian statistics (see Chapter 15) can prove useful in handling this kind of problem.

Another example of lack of homogeneity occurs when an auditor is checking inventory. Some types of inventory are more likely to be pilfered than others so the auditor would be wise to segregate out those types of inventory which are more likely to be pilfered and apply a more rigorous sampling check to these units.

A further and even more important example of a non-homogeneous population occurs in estimating sampling of values.

Most accounting populations such as inventory and debts have the characteristic that a few accounts represent a large proportion of the total value and many accounts are of insignificant value. Here again the auditor can achieve massive economies in sampling efficiency by segregating the units of high value from those of low value and applying a different sampling plan to each section.

This brings us to the important topic of *stratification*. But before we explore this topic further let us take a brief look at another statistic called *skewness*.

## THE MEANING OF SKEWNESS

*Skewness* measures the symmetry or rather asymmetry of a population around the mean.

Figures 7.1 (a), (b) and (c) show the frequency distributions of three accounting populations. In Figure 7.1(a) the distribution is symmetrical around the mean value and has the rough shape of a bell.[1] This type of bell-shaped distribution is called a *Normal* distribution and it arises when a population of values is generated by a large number of independent factors. The height of college students or the length of leaves from a given tree will tend to form a Normal distribution like Figure 7.1(a).

Accounting and auditing populations are rarely found to form a Normal distribution. Most accounting and auditing populations are skewed like Figure 7.1(b). Figure 7.1(b) illustrates a population of creditors at a given point in time. A small number of suppliers account for a high proportion of the total debt owed. This is the usual situation found with accounting populations, they are skewed out to the *right*.

Figure 7.1(c) shows a population skewed out to the left. This is a population of the months that have elapsed before legal action is taken against late paying customers. Populations skewed to the left are rarely found in accounting and auditing work.

Since most accounting populations are skew and many are highly skew it is important for an auditor or accountant to calculate the skewness inherent in an accounting population.

Perhaps we should also mention at this point that many statistical tables are calculated on the assumption that the population under scrutiny is Normally distributed around the mean. If this is not so, the tables will give misleading sample sizes. We shall return to this point when we discuss 'one tailed' distributions.

We must, then, be careful not to use tables of the Normal distribution to make predictions about highly skewed distributions. However, methods are available for 'normalizing' skewed distributions by 'chopping off' the long tail, i.e. stratifying the population as mentioned earlier or by taking the log of the various readings.

The latter statistical trick is much simplified by using logarithmic ruled graph paper.

[1] All symmetrical distributions are not, of course, normal.

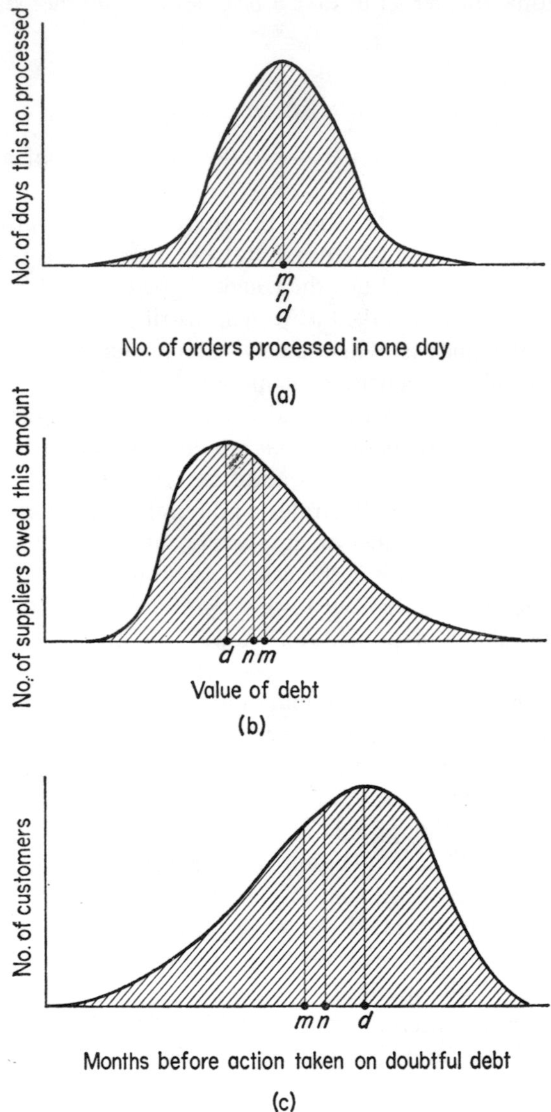


Fig. 7.1. Three accounting populations. (a) A symmetrical distribution. (b) A distribution skewed to the right. (c) A distribution skewed to the left.

### HOW DO WE MEASURE THE SKEWNESS OF A POPULATION?

We have already explained in Chapter 3 how to measure the mean of a population. Now the mean of a population is obviously affected by the value of a few very large amounts such as we find in a skewed population. The mean will be pulled out towards the right if the population is skewed to the right.

The *mode* of a population is the most common value in the population. If the population is *unimodal*, i.e. has a single hump, the mode is located by dropping a line from the highest point on the curve to the $x$ axis below as in Figure 7.1.

It is obvious that the mode of a population is not affected by the presence of extreme values in the population.

These two facts, the first that the mean of a population *is* affected by extreme values and that the mode is not, can be used to measure the skewness of a population.

The relevant formula is:

$$\text{skewness} = \frac{m - d}{s}$$

where $m$ = mean, $d$ = mode, $s$ = standard deviation.

For example for the data in Table 5.1 on p. 37 the calculation of the skewness is as follows:

| | (a) | (b) |
|---|---|---|
| mean = | 9·92 | 9·92 |
| mode = | 10 | 10 |
| standard deviation = | 1·63 | 2·67 |
| skewness = | $\frac{9{\cdot}92 - 10}{1{\cdot}63} = -0{\cdot}05$ | $\frac{9{\cdot}92 - 10}{2{\cdot}67} = -0{\cdot}03$ |

The skewness is negligible in both cases.

The populations are symmetrically distributed around the mean.

If however we take the data from the population set out in Table 7.1 on page 68 we arrive at a very different measure of skewness.

mean = £250
mode = £100 (say)
standard deviation = £360

$$\text{skewness} = \frac{250 - 100}{360} = +0{\cdot}42$$

a substantial volume of skewness. Note that a + value of skewness means that the population is skewed to the right and a − value that it is skewed to the left.

## USING THE MEDIAN INSTEAD OF THE MODE

In the last example we stated that the mode, or most common reading, was £100. In practice it is often difficult to calculate the mode of an accounting population and statisticians prefer to use the median.

The median, like the mode is not affected by the *size* of extreme values in the population.

Karl Pearson showed that the median falls roughly two thirds of the way from the mode to the mean and used this fact to devise the formula

$$\text{skewness} = \frac{3(m - n)}{s}$$

where $m$ = mean, $n$ = median, $s$ = standard deviation.

Returning to the examples given above, the skewness for the examples in Table 5.1. are

| | (a) | (b) |
|---|---|---|
| skewness = | $\frac{3(9{\cdot}92 - 10)}{1{\cdot}63}$ | $\frac{3(9{\cdot}92 - 10)}{2{\cdot}67}$ |
| | $= -0{\cdot}15$ | $= -0{\cdot}09.$ |

The distributions are so symmetrical that the mode and median coincide. The relative measure of skewness is therefore unaffected by substituting median for mode.

For the example given in Table 7.1 on p. 68 the new measure of skewness is

$$\text{skewness} = \frac{3(250 - 138)}{360} = +0{\cdot}93$$

We *estimate* the median in this example in the following way.

If we assume that the readings are spread equally across the class interval £0 to £200 then the 1000th or middle reading must lie at point 1000/1450 × £200 across the interval.

$$\frac{1000}{1450} \times £200 = £138$$

This rough approximation of the median is sufficient for most auditing and accounting purposes.

In this case the measure of skewness can vary from $+3$, extreme skew to the right, to $-3$, extreme skew to the left. It is rare to find an accounting population with a skewness greater than $\pm 2$, and we have already stated that few accounting populations are skewed to the left.

## MODE OR MEDIAN?

Should we use the mode or median to estimate skewness?

It all depends whether or not the population is segregated into classes as in Table 5.1 on p. 37. If it is so segregated clearly it is simpler to estimate the median. If it is not so segregated it may be easier to estimate the mode by taking a random sample of about 100 units from the population.

The mode can be connected to the median by substituting in the formula:

$$\text{median} = \frac{\text{mode} + (2 \times \text{mean})}{3}$$

Note that the latter formula becomes inaccurate if the skewness, based on the median, exceeds 2; this is rare.

## THE PROBLEM OF STRATIFICATION

The sample size required by an auditor to audit *values* can be massively reduced by the intelligent use of stratification. In many cases the use of statistical sampling methods to audit values such as debtors or inventory, is just not viable unless some form of stratification is used.

The following section will mainly consider the audit of values. In Chapter 11 we will make a brief comment on the problem of stratifying populations when the auditor is testing proportions.

## STRATIFYING DEBTORS, INVENTORY AND OTHER VALUE POPULATIONS

If the reader turns to Table 5.3 he will see that standard deviation is a very important determinant of sample size. With population size and level of confidence given, sample size, in estimation sampling of variables, is determined by the ratio; unit precision limit/standard deviation.

Now the unit precision limit tells us how precise an estimate is required from the sample and its value depends upon the estimated mean of the population.

The ratio of standard deviation/mean is a very important one in statistics and it has been given a special name. It is called the coefficient of variation.

Where the coefficient of variation is less than 0·5 it is likely that audit samples of around conventional size will be sufficient to test the population, but when the coefficient of variation exceeds 0·5 and approaches 1·0 it is likely that some form of stratification will be needed. When it exceeds 1·0 stratification is essential. The coefficient of variation of most unstratified accounting populations exceeds one.

Let us take an example.

| | |
|---|---|
| Population | 10,000 |
| Precision required | ±3% |
| Expected mean value | £10 |
| Confidence level | 90% |

| Population | SD | Mean | C of V |
|---|---|---|---|
| A | 5 | 10 | 0·5 |
| B | 10 | 10 | 1·0 |
| C | 15 | 10 | 1·5 |

The precision limit and sample size is therefore as follows.

| Population | Unit prec. limit/SD | | Required sample size |
|---|---|---|---|
| A | 0·3/5 | 0·06 | 700 |
| B | 0·3/10 | 0·03 | 2300 |
| C | 0·3/15 | 0·02 | 4000 |

Clearly the sample sizes for C of V above 0·5 are not viable and so the auditor must consider stratifying the population by value.

## THE OPTIMAL STRATIFICATION PLAN

Since the sample size depends upon the value of the standard deviation of the strata the optimal stratification plan is one which minimizes the standard deviation of the stratum. That is to say it reduces the variability of the values within the stratum to a minimum. The more strata that are created the more homogeneous is each stratum and so the lower will be the average standard deviation of all the strata.

However, the cost of sampling increases as the number of strata is increased since the auditor can no longer draw a simple unrestricted random sample but must draw *n* units from each stratum, and each stratum in the population may not be identified in advance for the auditor.

There is thus a trade-off between drawing cheap large samples from relatively undifferentiated populations and drawing small expensive samples from highly differentiated populations.

The auditor will probably compromise by segregating the population into either two or three strata.

Once the auditor has decided on the number of strata he must decide on the cut-off points for each stratum. For example if a population of debts runs from £1 to £100,000 he may decide to use the cut-off points

| Stratum | (a) Value | (b) Value | (c) Value |
|---|---|---|---|
| A | 0–999 | 0–99 | 0–9999 |
| B | 1000–9999 | 100–499 | 10,000–49999 |
| C | 10,000 and above | 500 and above | 50,000 and above |

Whichever of these schemes he adopts depends upon the distribution of value in the population. If a Lorentz curve (see ahead) or its tabular equivalent is available for a previous year this is helpful in choosing cut-off points which allocate roughly an equal proportion of the total value to each stratum. However we must not forget that the object of the exercise is to minimize the variability within each stratum and the strategy of allocating equal value may not achieve this.

Once the number of strata and the cut-off points are decided upon, a very simple rule can be used to allocate a fixed total audit sample between the strata to achieve maximum efficiency of sampling. This rule is: allocate the total sample to each stratum

$$\frac{\text{No. of units in the stratum} \times \text{SD of stratum}}{\text{Sum total of above for all strata}}$$

Suppose, for example, that an auditor decides to draw a sample from a population consisting of 100,000 lots of inventory. The inventory lots vary in value from £1 to £10,000. He decides on three strata with the following cut-off points.

| Stratum | Value £ | SD £ | No. of units | SD × No. of Un. | Sample size % |
|---|---|---|---|---|---|
| A | 0–99 | 60 | 80,000 | 4,800,000 | 33·5% |
| B | 100–999 | 300 | 15,000 | 4,500,000 | 31·5% |
| C | 1000 and over | 1000 | 5000 | 5,000,000 | 35·0% |
| | | | | | 100·0% |

Whatever the fixed sample size, it ought to be divided between the strata in the proportion calculated above. This allows the auditor to derive the maximum information from the fixed sample size.

This method of deciding on the number of units to draw from each stratum is relatively easy to follow. But the auditor does not normally decide on a fixed sample size in advance. He usually decides on the confidence level and precision limit he needs in his inference and from these he calculates the required sample size.

In any case, in the previous example he would need to calculate the precision limits on his estimate.

One method of calculating the sample size required in stratified sampling is as follows.

### EXAMPLE OF STRATIFICATION

Suppose that a population of debtors can be stratified as follows.

Table 7.1. Population of debts stratified.

| Stratum | Value | Number | Percentage | Sample estimate of average value of stratum £ | Standard deviation of stratum (estimate) £ |
|---|---|---|---|---|---|
| A | 0–200 | 1450 | 72·5 | 100 | 20 |
| B | 201–1000 | 500 | 25·0 | 500 | 70 |
| C | over 1000 | 50 | 2·5 | 2000 | 1000 |
| Total population | | 2000 | 100 | 250 | 360 |

Suppose we do *not* stratify this population and wish to estimate the value of the population to within ±£20,000 at a 90% level of confidence. What sample size do we need?

The answer from Figure 21.5.3 on p. 251 turns out to be a sample size of over 1000 units.

If however we use the stratified sample set out in Table 7.1 what sample size do we need?

First we must calculate the *stratification coefficient*. This is calculated using the following formula.

$$t = \frac{\sum(ns)}{N^2(cp)^2 + \sum(ns^2)}$$

where $t$ = stratification coefficient; $n$ = the number of units in each strata; $s$ = the standard deviation of each strata; $p$ = required unit precision limit; $N$ = population size; $c = 0{\cdot}606$ for confidence level of 90%, 0·51 for confidence level of 95%, 0·388 for confidence level of 99%.

Let us apply this formula to the problem set out in Table 7.1 above.

Since stratum C is so small and accounts for such a large part of the dispersion we decide to carry out a 100% check on the stratum.

We have eliminated all sampling error from stratum C and so we now concentrate on strata A and B.

We can set out the data on strata A and B as follows:

| Stratum | No. of units ($n$) | SD ($s$) | $n \times s$ | $n \times s^2$ |
|---|---|---|---|---|
| A | 1450 | 20 | 29,000 | 580,000 |
| B | 500 | 70 | 35,000 | 2,450,000 |
| | 1950 | | 64,000 | 3,030,000 |

Therefore the stratification coefficient is calculated as follows:

$$t = \frac{64{,}000}{3{,}802{,}500 \times (0{\cdot}606 \times 10)^2 + 3{,}030{,}000} = \frac{64{,}000}{143{,}000{,}000} = 0{\cdot}000447$$

We now multiply together, for each stratum, number of units in stratum × standard deviation of stratum × stratification coefficient. The answer to this sum is the required sample size for each stratum.

Applying the above to the problem set out in Table 7.1 above:

| Stratum | $n$ | $s$ | $t$ |
|---|---|---|---|
| A | 1450 | 20 | 0·000447 = 13 |
| B | 500 | 70 | 0·000447 = 15 |
| C | 100% sample | | = 50 |
| Total sample size | | | 78 |

By stratifying the population we have reduced the required sample size from over 1000 to 78.

Let us admit immediately that stratification does not always produce a reduction in sample size as dramatic as this. But the example illustrates what is possible. The large standard deviation of £360 was a consequence of the large variance in stratum C. By including all of stratum C in the sample we abolished all sampling error from this source and reduced the standard deviation for the remainder of the population to around £25. This, in turn, resulted in the massive reduction in sample size.

*Stratification is, therefore, a subject of great importance to the auditor who wishes to use statistical sampling for testing values.*

By stratifying a population an auditor is very likely to reduce sample size considerably. Usually by at least 50%. He is most unlikely to increase sample size by using the technique.

The selection of optimal cut-off points is a skill that comes with experience. The basic principle is to draw a larger sample from the strata with the higher *absolute* dispersion of values.

The optimal degree of stratification is achieved when the strata averages are as far apart as possible and the standard deviations of each stratum are as small as possible.

The optimal arrangement occurs when the sampling fraction in each stratum is proportional to the standard deviation of the stratum times the number of units in the stratum. The formulae given above approximates to this ideal.

Accounting populations are so large and complex that an auditor can never hope to reach the ideal stratification, but by striving towards it he can greatly improve the information he derives from a given size of sample.

I again repeat that an auditor cannot hope to make much use of estimation sampling of variables unless he has a clear understanding of the principles and procedures of stratification.

## STRATIFICATION IN PRACTICE

Once a population is stratified the auditor will have little trouble in drawing a random sample from each stratum. But most accounting populations are *not* stratified by value, therefore the auditor has the initial problem of segregating out the sample frame from which he will draw his sample.

Once the auditor has decided on his cut-off points he may be able to persuade his client to keep his ledgers in $n$ separate sections, each section representing one stratum. This is easy to achieve if the accounts, etc., are kept on separate ledger cards.

Another solution is to 'flag' cards within each stratum with a different coloured tag.

If a list of debts, etc., is available a systematic random sample can be drawn from each stratum by using the following procedure.

1. Estimate the number of units in each stratum using a preliminary sample.
2. Calculate the number of units to be drawn from each stratum.
3. If $s$ units have to be drawn from a stratum of size $N$ a systematic sample will select every $N/s$ unit in the stratum, i.e. if we wish to select

500 units from a stratum containing 5000 units we select every tenth unit starting at a randomly selected point among the first ten.

If for example debts are listed by value and the stratum is £100 to £500 we select every tenth unit *in this value range* and ignore the other debts which are included in other strata.

A stratum random sample can be drawn relatively easily by using one or other of these methods.

If the population being audited is summed on a regular basis by a comptometer operator the operator can be trained to select out a random sample using the following method.

1. Estimate the total value of the population, say £100,000.
2. Calculate the total required sample size, say 200 units.
3. Select 200 random numbers between 1 and 99,999.
4. As the operator sums the list of items she is trained to tick any value which contains one of those random numbers.

This method provides a sample with the probability of selection proportional to the value of the item.

The method will oversample from units of high value. Later the auditor will need to select an additional sample to top up each stratum sample to the required size. However, a good portion of the sample will have been drawn by the comptometer operator prior to the auditor's arrival.

The MUS system described in Chapter 17 uses a variant of this method for selecting a sample. The method overcomes the skewness problem by sampling £s rather than accounts for testing.

Let us take a simple example.

Population 176, 234, 973, 45, 633, 986, 8, 554, 6390.
Total 9999
Sample size required 3 units
Random numbers selected 3854, 334, 7176, 1648.
Cumulative value of population.

| | |
|---|---|
| 000–176 | |
| 177–410 | (334) |
| 411–1383 | |
| 1384–1428 | |
| 1429–2061 | (1648) |
| 2062–3047 | |
| 3048–3055 | |
| 3056–3609 | |
| 3610–9999 | (3854)(7176) |

The 2nd, 5th and 9th random units were selected. If two random numbers fall within the same number ignore all but the first since the auditor samples without replacement. This introduces a slight bias into the sample but it is not of large magnitude.

## OTHER STRATEGIES FOR STRATIFICATION

It often happens that, as in the example given, the top stratum can be 100% sampled. This may reduce the standard deviation to such an extent that a simple random sample can be drawn from the rest of the population. Thus the complex stratification formula set out above need not be resorted to.

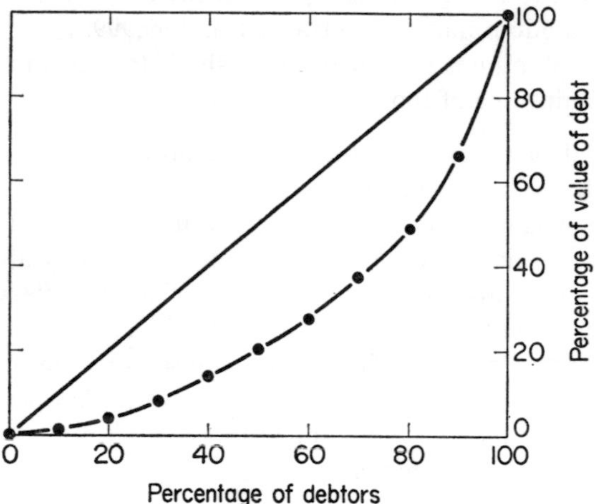


Fig. 7.2. Lorentz curve of distribution of total value of debt among debtors.

| Cumulative percentage of debtors | Cumulative percentage of value of debt |
|---|---|
| 10 | 1 |
| 20 | 4 |
| 30 | 9 |
| 40 | 14 |
| 50 | 20 |
| 60 | 28 |
| 70 | 38 |
| 80 | 49 |
| 90 | 65 |
| 100 | 100 |

## THE LORENTZ CURVE

Earlier in this section we mentioned the use of the Lorentz curve to estimate the distribution of value among the units of a population.

A typical Lorentz curve is shown in Figure 7.2. This illustrates the cumulative percentage of value owed by the cumulative percentage of debtors. The tabular representation of the Lorentz curve shown below the figure is the form most likely to appeal to the auditor.

The Lorentz curve is useful for deciding on stratum cut-off points. If no supplementary information is available the auditor, as a first approximation, can choose cut-off points which will allocate roughly $100/n\%$ of value to each stratum, if there are $n$ strata.

With the population illustrated in Figure 7.2 a reasonable stratification procedure would be three strata, the first strata accounting for the first 70% of debtors, the next strata accounting for the next 20% and the third strata accounting for the remaining, high value, 10%. Thus each stratum accounts for roughly one third of the total value.

## QUESTION SERIES 7

1. What is an attribute of a unit of a population? State three attributes of a credit account.
2. What is the name of the statistical technique which estimates the proportion of a population having a given attribute?
3. Give three examples of estimation sampling of variables.
4. In checking month end statements, the auditor tests invoice codes and amounts, cash payments, summation, customer code and date. How many populations is he checking? (*)
5. What do we mean when we say that a population is *homogeneous*?
6. If a population is *not* homogeneous what should an auditor do?
7. Give an auditing example where stratification might prove useful. (*)
8. What procedure would you follow in stratifying a debtors population by value?
9. Why does a stratified sample produce more information than an unrestricted sample of the same size? (*)
10. Is a Normal distribution another word for any distribution that is symmetrically dispersed around the mean value? (*)
11. Draw a distribution that is skewed out to the right.

12. Can you think of another accounting distribution that is skewed out to the left?
13. Estimate the skewness of the data in Table 5.1 on p. 37
14. If the skewness of an accounting population you are auditing comes out at 0·9 what does this tell you? (*)

Table 7.2. One hundred readings of an accounting variable. The variable might be debts, inventory, error value, etc.

| Row \ Column | 1/2 | 3/4 | 5/6 | 7/8 | 9/10 |
|---|---|---|---|---|---|
| 1 | 82 | 71 | 128 | 83 | 97 |
| 2 | 63 | 89 | 78 | 129 | 78 |
| 3 | 90 | 131 | 139 | 71 | 89 |
| 4 | 78 | 92 | 87 | 61 | 125 |
| 5 | 123 | 158 | 92 | 83 | 119 |
| 6 | 110 | 83 | 82 | 102 | 74 |
| 7 | 87 | 57 | 101 | 93 | 85 |
| 8 | 135 | 109 | 60 | 94 | 107 |
| 9 | 96 | 75 | 76 | 99 | 91 |
| 10 | 76 | 64 | 112 | 52 | 68 |
| 11 | 62 | 95 | 98 | 78 | 105 |
| 12 | 54 | 119 | 83 | 114 | 83 |
| 13 | 117 | 129 | 102 | 106 | 80 |
| 14 | 101 | 84 | 108 | 84 | 115 |
| 15 | 91 | 74 | 114 | 130 | 71 |
| 16 | 86 | 100 | 76 | 123 | 94 |
| 17 | 125 | 97 | 87 | 141 | 86 |
| 18 | 70 | 80 | 96 | 82 | 77 |
| 19 | 106 | 110 | 87 | 73 | 68 |
| 20 | 86 | 145 | 155 | 100 | 149 |

15. One hundred readings of an accounting population are set out in Table 7.2. Calculate or prepare the following for this population:
    (a) Frequency distribution and histogram based on class values of £50/59, £60/69, etc.
    (b) Mean and standard deviation of each column and of total population.
    (c) Median, mode and skewness of total population.
    (d) Lorentz curve of distribution of value of debts among debtors, assuming these readings represent debtors balances. Use 10% blocks. (*)

16.

| Stratum | No. of units | Percentage | Sample estimate of average value of stratum £ | Estimated SD of stratum £ |
|---|---|---|---|---|
| 0–50 | 700 | 70 | 1 | 0·5 |
| 51–100 | 200 | 20 | 5 | 3·0 |
| 101–500 | 100 | 10 | 10 | 4·0 |
| Total population | 1000 | 100 | 2·7 | 2·0 |

An auditor wishes to verify the value of a population of inventory consisting of 1000 types of stock. A preliminary sample of 100 provides the information set out above.

If we suppose that the auditor wishes to estimate the value of inventory to within a unit precision limit of £0·14 at a 95% level of confidence, calculate:

(a) Sample size required without stratification.
(b) Sample size required with stratification.
(c) What percentage reduction in sample size has stratification achieved?

*Note*: Do not sample any stratum 100%.

(d) If the auditor decides to draw a block sample of 50 units from this population, how should he divide the sample between the strata? (*)

## SOME ANSWERS TO QUESTION SERIES 7

4. Six.
7. Estimating error rate. Staff of different departments known to have varying clerical abilities on the average in each department. Stratify high from low expected error rate.
9. Standard deviation—on average—reduced.
10. No. The Normal is a special kind of symmetric distribution.
14. It is rather highly skewed to the right. Therefore it is probable that a few accounts represent a high proportion of the total value.

15.

Table 7.3. Frequency distribution of the 100 readings of an accounting variable set out in Table 7.2. Note that actual mean is 95·2, but that the mean calculated from mid-point of class interval—95·7—is very close to this.

$$\text{Standard deviation} = \left(\frac{51{,}900}{99}\right)^{1/2} = 22{\cdot}90$$

Mean (based on midpoint of class interval) = 95·7.

| (a) Value | (b) Frequency | (c) Deviation from mean | (d) (Deviation)$^2$ | (e) (b) × (d) |
|---|---|---|---|---|
| 50–59 | 3 | −40 | 1600 | 4800 |
| 60–69 | 7 | −30 | 900 | 6300 |
| 70–79 | 16 | −20 | 400 | 6400 |
| 80–89 | 22 | −10 | 100 | 2200 |
| 90–99 | 15 | 0 | 0 | 0 |
| 100–109 | 12 | +10 | 100 | 1200 |
| 110–119 | 9 | +20 | 400 | 3600 |
| 120–129 | 7 | +30 | 900 | 6300 |
| 130–139 | 4 | +40 | 1600 | 6400 |
| 140–149 | 3 | +50 | 2500 | 7500 |
| 150–159 | 2 | +60 | 3600 | 7200 |
| | 100 | | | 51,900 |

16. (a) Without stratification the required sample size is 363.
(b) With stratification.

| Stratum | $n$ | $s$ | $n \times s$ | $n \times s^2$ |
|---|---|---|---|---|
| A | 700 | 0·5 | 350 | 175 |
| B | 200 | 3·0 | 600 | 1800 |
| C | 100 | 4·0 | 400 | 1600 |
| | 1000 | | 1350 | 3575 |

The stratification coefficient is therefore:

$$t = \frac{1350}{1{,}000{,}000 \times (0{\cdot}51 \times 0{\cdot}14)^2 + 3575} = \frac{1350}{8672} = 0{\cdot}1557$$

The sample required from each stratum is therefore:

| Stratum | $n$ | $s$ | $t$ | Sample size |
|---|---|---|---|---|
| A | 700 × | 0·5 × | 0·1557 = | 55 |
| B | 200 × | 3·0 × | 0·1557 = | 94 |
| C | 100 × | 4·0 × | 0·1557 = | 63 |
| Total sample | | | | 212 |

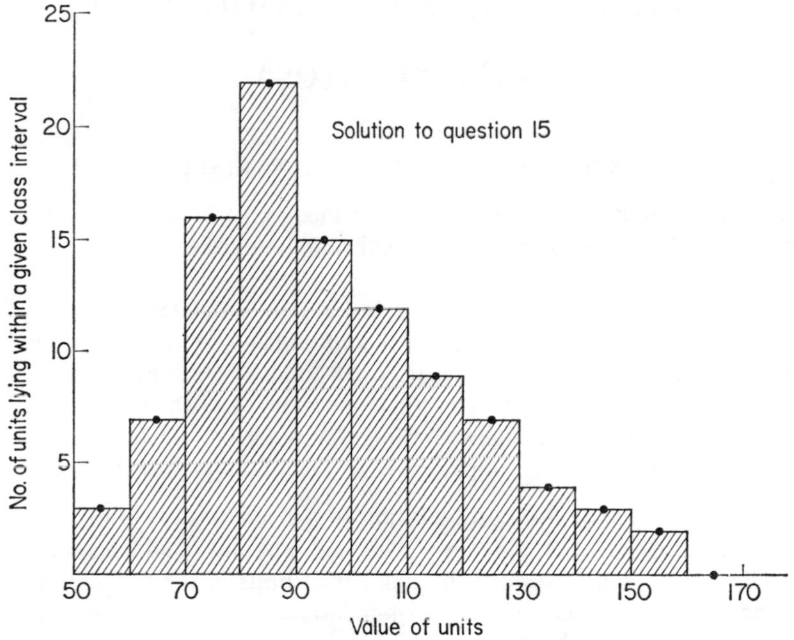


Fig. 7.3. A frequency histogram describing the 100 units from an accounting population set out in Table 7.2. Note that the distribution is slightly skewed to the right.

(c) By stratifying the population the auditor has reduced sample size by 42%.

(d)

| Stratum | Percentage | No. |
|---|---|---|
| A | 26 | 13 |
| B | 44 | 22 |
| C | 30 | 15 |
| | 100 | 50 |

# 8

# Estimation Sampling of Values (Variables) and Proportions (Attributes)

## ESTIMATION SAMPLING OF VARIABLES

Estimation sampling of variables is the most general form of statistical sampling. All other forms of statistical sampling are special cases of this general form.

Estimation sampling of variables is used to estimate the *value* of a population. The most common uses in auditing and control are estimating the value of assets, such as debtors, small tools and inventory, the value of liabilities, such as trade or expense creditors, and the value of error.

The procedure for using estimating sampling of variables has been explained previously. A summary of the various steps is as follows:

1. The auditor, after due consideration of the circumstances of the case, must decide on a suitable level of confidence and confidence interval.
2. The auditor must estimate the number of units in the population and the standard deviation of the population. He may, at this point, decide to stratify the population.
3. Since all of the determinants of sample size are known the auditor can now calculate the sample size needed to meet the various conditions set out in steps (1) and (2) above.

   He *could* calculate sample size by substituting numbers for the symbols in the variables sample size formula. But it is quicker and possibly more accurate, to consult one of the several tables or graphs of estimation sampling of variables, such as that set out on pp. 250–256.
4. A required sample of size $p$ has been calculated. The auditor must now select $p$ random units from the population of $P$ units. He can draw this sample using numbers from random number tables or by using one of the other methods set out in Chapter 6.
5. Once the random sample of $p$ units is drawn the auditor calculates the

sample mean, multiplies it by $P$, the number of units in the population, and arrives at an estimate of £$V$, the value of the population.

Since the confidence interval $\pm$ £$c$ at an $x\%$ level of confidence is known the auditor can now state that he is $x\%$ confident that the value of the population of $P$ units lies between £$(V - c)$ and £$(V + c)$.

Alternatively if the total value, £$y$, of the population is given, and the auditor is testing the validity of this value, he tests to see whether the value £$v$, extrapolated from the sample, falls within the confidence interval £$(y - c)$ to £$(y + c)$. If $v$ does fall in this interval he accepts £$y$ as the value of the population and assumes £$(y - v)$ to be caused by sampling error. If £$v$ falls outside the confidence interval the auditor either rejects £$y$ as the value of the population or increases his sample size to carry out a more rigorous test.

Let us take an example.

| | |
|---|---|
| Population size | 2000 |
| Standard deviation of population | £2 |
| Level of confidence (two tailed) | 90% |
| Unit precision limit | ± £0·2 |
| Sample size | ? |

*Solution*

1. First let us calculate the ratio:

$$\frac{\text{unit precision limit}}{\text{standard deviation}} = \frac{0{\cdot}2}{2{\cdot}0} = 0{\cdot}10$$

2. We look up graph on p. 251 and find that for the given conditions the required sample size is close to 240 units.
3. We draw 240 random numbers and so 240 random units.
4. The total value of the sample of 240 units turns out to be £2520. We divide this figure by 240 to arrive at a mean value of £10·5.
5. We now multiple £10·5 by 2000, the number of units in the population, to arrive at an estimate of £21,000 for the value of the population.
6. The unit confidence limit is ± £0·2, so the population confidence limit is $(2000 \times 0{\cdot}2) =$ £400.

Our first estimate of the value of the population is then £21,000 ± £400, or in other words that the value of the population lies between £20,600 and £21,400 with a 90% level of confidence.

But we may be able to improve the accuracy of our estimate still further.

Suppose that the population we are testing is inventory. The auditor is not worried if inventory exceeds £21,400 but only if it is less than £20,600. The auditor is not interested in the 5% of possible values exceeding £21,400.

This being the case the auditor can improve the accuracy of his inference by stating 'I am 95% confident that the value of the population is not less than £20,600'.

This is a considerable improvement on the original inference 'I am 90% confident that the value of the population lies between £20,600 and £21,400'.

We will return to the question of 'one tailed inferences' in Chapter 11.

If the auditor decides to stratify the population into several homogeneous strata he can effect an even greater improvement in the accuracy of his inference.

## CALCULATING SAMPLE SIZE USING FORMULAE

If suitable tables are not available the auditor can calculate the required sample size in variables sampling by using the formula:

$$n = \frac{(S)^2}{(s)^2} \tag{1}$$

where $n$ = required samples size, $S$ = standard deviation of population, $s$ = standard error (deviation) of sample means.

The reader will recall from Chapter 3 that the standard error of the sample mean was related to the standard deviation of the population as follows:

$$s = \frac{S}{(n)^{1/2}} \tag{2}$$

Formula (1) above is simply formula (2) expressed in terms of $n$, the sample size.

Returning to our problem we see that we know $S$, the standard deviation of the population, but how do we calculate $s$, the standard error of the sample mean?

We calculate $s$ as follows:

1. The auditor has stated that he requires a 90% level of confidence that the actual unit population mean lies within ±£0·2 of the estimated unit population mean.
2. Since the sample means are normally distributed we know that the

actual mean lies $\pm 1{\cdot}64$ standard errors from the estimated mean 90% of the time.[1]

3. Therefore, 1·64 standard errors must equal £0·2 if the conditions set are to be satisfied. Thus

$$1{\cdot}64s = 0{\cdot}2$$
$$s = 0{\cdot}122$$

The standard error of sample means is £0·122.

Substituting in formula (1) we find:

$$n = \frac{(2)^2}{(0{\cdot}122)^2} = \frac{4}{0{\cdot}0149} = 268$$

This is rather larger than our original estimate of 240. The reason is that we have assumed our population to be of infinite size. The graph makes allowance for the population being a finite population of 2000 units.

The formula for finite populations is;

$$n = \frac{S^2}{s^2 + S^2/N}$$

where $n$ = size of sample, $S$ = estimated standard deviation of the population, $s$ = calculated standard deviation of the sample means, $N$ = size of the population.

In the previous example,

$$S = 2$$
$$s = 0{\cdot}2/1{\cdot}64 = 0{\cdot}122$$
$$N = 2000$$

therefore,

$$n = \frac{4}{0{\cdot}0149 + 0{\cdot}002} = 237 \text{ units.}$$

Compare this to the result provided in Figure 21.5.3.

## ESTIMATING THE VALUE OF DIFFERENCES

A useful rule in sampling is the maxim 'never waste information'. Any prior information you have about a population can almost invariably be used to squeeze more information out of a given sample size or to reduce the required sample size.

[1] See Table 21.1

A good example of this occurs when an auditor is attempting to verify a total made up of a large number of individual units. In the previous examples in this chapter the auditor did not use the total except to verify that it fell within the projected confidence limits. If, instead of adopting this approach he uses the total as the base from which to calculate error differences he may be able to effect a reduction in sample size even more dramatic than that effected by stratified sampling.

Let us work through an example.

A large part of a company's costs consists of the cost of carriage and transportation. The company owns no transportation facilities itself but uses outside hauliers.

The calculation of transport rates is a complicated procedure and for several years the company accountant has simply checked that month by month the total transport bill seems reasonable.

In 19?5 the chief accountant is shocked to read about a case where a company has been grossly overcharged for transport over several years because they had failed to check the rates. He determines to rectify this omission in his own case.

The accountant could test, say, 200 of the 10,000 transport vouchers, calculate the corrected mean value of each and extrapolate these corrected amounts to be compared. For example he knows that the total transportation charge for the year was £832,700. He wishes to check that this amount falls within 2% of the amount he should have been charged, that is within approximately £17,000.

What sample size does the auditor need to achieve this accuracy with a 90% level of confidence? See Table 8.1.

He draws a preliminary sample of 98 vouchers and estimates the standard deviation to be £41.

The relevant statistics are therefore

| | |
|---|---|
| Population size | 10,000 |
| Confidence level | 90% |
| Precision limit | £17,000(0·85) |
| Unit precision limit | £1·7 |
| Standard deviation | £41 |
| UPL/SD | 0·04 |

Figure 21.5.5 on p. 253 suggests that the required sample size from the *corrected* vouchers is so large that it is quite obviously uneconomic. Using the formula it turns out to be around 1400 vouchers! To reduce sample size to around 200 we would have to widen our precision limit to about £45,000 or over ±5%. However if we concentrate our attention on the

Table 8.1. Comparison of invoiced value of transportation vouchers with value derived from precise calculation. No. of vouchers, 10,000; percentage of sample of 100 which differ, 10% (approx.); average difference on sample of 200, +£1·2.

| | Invoiced value £ | Checked value £ | Difference |
|---|---|---|---|
| | 127 | 127 | — |
| | 94 | 93 | +1 |
| | 36 | 36 | — |
| | 221 | 224 | −3 |
| | 73 | 69 | +4 |
| | 68 | 68 | — |
| | 81 | 81 | — |
| | 112 | 109 | +3 |
| | 13 | 13 | — |
| | · | | |
| | · | | |
| | · | | |
| | 96 | 96 | — |
| Total | 16,944 | 16,704 | +240 |
| Average | 84·72 | 83·52 | +1·2 |

*differences* between the original and corrected vouchers we can achieve a much narrower precision limit at the same sample size.

To begin with we see that 90% of the initial random sample of 100 are correct. From this fact we can infer that we are 95% confident that the percentage of incorrect vouchers is less than 16%—see Figure 21.2.1. However this is not the main purpose of the test.

We draw a further 100 vouchers and find the average value of the difference between the unchecked and checked vouchers to be +£1·2. That is, on the average, the unchecked vouchers are too high by £1·2 per voucher. These differences range from −£28·1 to +£116·4 with a standard deviation of £16·8. What sample size do we need now to calculate the total difference to within £17,000?

The relevant facts are:

| | |
|---|---|
| Population size | 10,000 |
| Confidence level | 90% |
| Unit precision level | £1·7 |
| Standard deviation | £16·8 |
| UPL/SD | 0·10 |

The sample size from Figure 21.5.5 turns out to be around 260, a massive reduction on 1500. Yet we have achieved the same answer!

If the reader compares the two sets of figures he will see that the only difference lies in the figure for standard deviation. The value of SD of the differences is only 41% of the SD of the corrected values.

Returning to our examples. We draw a further 60 vouchers and the average value of the 260 differences turns out to be +£1·3. We multiply £1·3 by 10,000 to arrive at the mean estimate of difference for the population. This is £13,000. We are therefore 90% sure that the correct value of the total of the transport vouchers lies between £(832,700 − 13,000 − 17,000) and £(832,700 − 13,000 + 17,000). That is between £802,700 and £836,700. We are also around 94% sure that the correct value is less than the original charge of £832,700.

The accountant is now in a position to decide whether the incremental cost of a more detailed check on transport charges is justified by the likely saving. In many auditing situations a comparison is made between two sets of figures. In such situations the auditor should always check whether the 'difference' approach can be used. If it can a large economy in sampling size is likely to be effected.

## RATIO ESTIMATES

An alternative method of increasing accuracy of inference without increasing sample size is to use the ratio estimate method. This method is rather similar to the difference approach. Returning to the example of the previous section, we draw a random sample of 200 units and find the sample totals to be as follows.

| | |
|---|---|
| Total value of unchecked sample | £16,944 |
| Total value of checked sample | £16,704 |

The ratio of these two figures to one another is 0·9858.

To estimate what the transport charge ought to have been we multiply the actual charge £832,700 by 0·9858 which gives us £820,876. But how accurate is this estimate, in other words what are the confidence limits on this estimate? As usual we must estimate the standard deviation but in the present case we calculate the SD in rather an unusual way employing the following formula:

$$D = \left(\frac{\sum (c - r \times i)}{n}\right)^{1/2}$$

where $D$ = standard deviation, $c$ = checked value, $r$ = ratio, $i$ = invoiced value, $n$ = number of units in the sample.

A section of these calculations is shown in Table 8.2. The SD comes out to be £13·79. We require a unit confidence limit of £1·7 so the sample size we require is:

| | |
|---|---|
| Population | 10,000 |
| Confidence level | 90% |
| Unit precision limit | £1·7 |
| Standard deviation | £13·79 |
| UPL/SD | 0·12 |

Table 8.2. Calculation of standard deviation in estimating by ratio.

| (a) Checked value £ | (b) Invoiced value £ | (c) Ratio | (d) (b) × (c) | (e) (a) − (d) | (f) (e)² |
|---|---|---|---|---|---|
| 127 | 127 | 0·9858 | 125·19 | −1·81 | 3·28 |
| 93 | 94 | 0·9858 | 92·67 | −0·33 | 0·11 |
| 36 | 36 | 0·9858 | 35·49 | −0·51 | 0·26 |
| 224 | 221 | 0·9858 | 217·86 | −6·14 | 37·70 |
| 69 | 73 | 0·9858 | 71·96 | +2·96 | 8·76 |
| 68 | 68 | 0·9858 | 67·03 | −0·97 | 0·94 |
| . | . | . | . | . | . |
| . | . | . | . | . | . |
| . | . | . | . | . | . |
| 96 | 96 | 0·9858 | 94·64 | −1·36 | 1·85 |
| | | | | | 38,041·34 |

$$D = \left(\frac{\sum (c - r \times i)^2}{n}\right)^{1/2}$$

$$D = \left(\frac{38{,}041 \cdot 34}{200}\right)^{1/2} = £13 \cdot 79$$

From Figure 21.5.5. on p. 253 the sample size turns out to be about 180. A significant reduction on the sample of 260 required for the difference method. Since we already have a sample of 200 we will use these to make our estimate. The confidence limit on the total population estimate of £820,876 is 10,000 × £1·7 = £17,000.

Therefore we are 90% confident that the checked value of the transportation vouchers lies between £803,876 and £837,876.

## DIFFERENCE VERSUS RATIO ESTIMATES

The ratio approach provided an almost identical prediction to the difference approach yet it used a sample of 190 against 260. The reduction in sample size was caused by the ratio approach producing a smaller standard deviation. When the absolute value of the difference is highly correlated to the absolute value of the unchecked amount then the ratio method will be the more efficient method. That is, when a large difference is usually generated by a large value the ratio method will require a smaller sample size. This is the usual situation in accounting work.

## RATIO VERSUS SIMPLE RANDOM SAMPLING OF CHECKED POPULATION

As a general rule, if the correlation between the values of the unchecked and the checked population is high, ratio estimates will provide a more efficient sampling approach than simple random sampling of the checked population. The precise calculation is that if:

$$r > \frac{Sx \cdot mx}{2(Sy \cdot my)}$$

where $r$ = correlation coefficient between unchecked population $x$ and checked population $y$, $S$ = standard deviation, $m$ = mean.

Then the ratio method is more efficient than simple random sampling of the checked population. This is invariably the case in accounting work.

## A CAUTION

In the previous examples we have assumed that a reasonable number of differences occurred in the samples chosen. In order for the above methods to be reasonably accurate we need to find at least 30 differences in our sample. If large differences are usually related to large values the auditor may be able to stratify the population to segregate large differences relatively quickly.

## ESTIMATION SAMPLING OF ATTRIBUTES

Estimation sampling of variables measures the value of a population. Estimation sampling of attributes measures the *proportion* of a population having a given attribute.

Some auditing and control applications of ES of A are the proportion of:

Error in a population of vouchers.
Debts $n$ months overdue.
Inventory lots with a turnover less than $n$ months.
Employees receiving overtime payments during a given period.

With each of the above populations every unit in the population either has, or does not have, a given attribute. A voucher is either in error or it is not, a 'debt' is either $n$ months overdue or it is not. One part of the population, let us call this $p\%$, has the required condition, the remainder, $(1 - p)\%$ does not. Note that the two possible states of the unit are mutually exclusive.

The problem in ES of A is to estimate $p\%$, the proportion of a population having a given condition, by drawing a random sample of $n$ units from the population and inferring the population proportion from the sample proportion.

As with ES of V the making of an estimate is easy, the measuring of the accuracy of the estimate rather more difficult.

ES of A is, in fact, a special case of ES of V where the unit can have only one of two values 1 or 0. The value 1 means that the unit possesses the attribute, the value 0 that it does not possess the attribute.

Suppose we are attempting to measure the proportion of debts over three months or more overdue from the following population of months overdue:

1 7 3 2 1 1 2 1.

We can represent a debt three months or more overdue as 1 and a debt not three months overdue as 0. Thus the population becomes

0 1 1 0 0 0 0 0.

The mean of this population of values is $2/8 = 0{\cdot}25$. *Thus the proportion of the population having the required attribute*, 0·25 in this case, *is equal to the mean of the population.*

To find the standard error (deviation) of a sample estimate of this mean we use the formula:

$$s = \left(\frac{p(1 - p)}{n}\right)^{1/2} \tag{1}$$

where $p$ = proportion of sample having attribute, $n$ = size of sample.

This formula applies if the sample is small relative to the population, say less than 5%. When the sample makes up a relatively large proportion of the population we use the more accurate formula:

$$s = \left(\frac{p(1 - p)}{n}\right)^{1/2} \left(1 - \frac{n}{N}\right)^{1/2} \qquad (2)$$

where $N$ = size of population.

Note that in ES of V we had to estimate the mean *and* the standard deviation of the population. In ES of A, as formula (1) above shows, the standard deviation is estimated automatically once we know $p$, the proportion of the sample having the attribute. Therefore there is no need for an independent calculation of standard deviation with ES of A.

Let us take an example.

An auditor is testing 100,000 wage slips to estimate the proportion which includes overtime payments. He takes a random sample of 500 and finds 100 to include overtime payment. The proportion of wage slips including overtime payments lies between what two percentages with a 90% level of confidence?

Facts:

| | |
|---|---|
| Population | 100,000 |
| Sample | 500 |
| Level of confidence required | 90% |
| Percentage of sample with attribute | 20% |

*Solution*

Standard error of proportion is:

$$s = \left(\frac{20 \times 80}{500}\right)^{1/2} = (3{\cdot}2)^{1/2} = 1{\cdot}79$$

Now it can be shown that the sample means of the estimates of the population proportion are normally distributed around the population mean, unless the proportion is very small. Therefore

| We can be $x$% confident that | The sample mean lies within $n$ standard errors of the population mean |
|---|---|
| $x$% | $n$ |
| 90 | 1·64 |
| 95 | 1·96 |
| 99 | 2·58 |
| 99·7 | 2·97 |

In the problem given we were asked for a 90% level of confidence. Therefore the precision limits are:

$$20\% \pm (1{\cdot}79 \times 1{\cdot}64) = 20\% \pm 2{\cdot}935 \text{ (3\% approx.)}$$

The auditor can state with a 90% level of confidence that the percentage of payslips including overtime lies within the confidence interval 17% to 23%.

The above calculations are not difficult to make but as we explained above[1] the formula tends to become slightly inaccurate when the proportion being tested is less than 10%. Therefore the reader may prefer to consult a suitable table providing the precision limits on proportions under various conditions. Arkin (1963), appendix F, provides such a table.

In the above example we selected an arbitrary sample size and calculated the confidence interval associated with this sample size. The more usual situation is for the auditor to decide on confidence level and precision limit and then calculate the sample size to satisfy these requirements.

The formula for calculating sample size under these conditions is:

$$n = \frac{p(1-p)}{(e/f)^2 + [p(1-p)/N]} \qquad (3)$$

where $n$ = sample size; $p$ = an estimate of the proportion percentage; $N$ = number of units in population; $e$ = required precision limit; $f = 1{\cdot}64$ at 90% confidence level, 1·96 at 95% confidence level, 2·58 at 99% confidence level, 2·97 at 99·7% confidence level.

If we take the previous example and take sample size and not the precision limit (±3%) as the unknown, the formula (3) becomes:

$$n = \frac{(20 \times 80)}{(3/1{\cdot}64)^2 + (20 \times 80/100{,}000)} = \frac{1600}{3{\cdot}345 + 0{\cdot}016} = 476$$

The difference of (500 − 476) = 24 is caused by using the precision limit of ±3% rather than the precise limit of (1·79 × 1·64) = ±2·935%.

The reader may be troubled by the fact that in formula (3) the auditor must estimate *in advance* the likely answer to his question, 'what is the proportion $p$?'. This need present no problem so long as the auditor remembers that if his estimate is closer to 50% than the actual percentage the sample size will be a *conservative* estimate.

Moral: always estimate $p$ closer to 50% than you think it really is. But don't exaggerate or you will finish up with a sample which is much too large. For example if $p$ is valued at 30% rather than 20% in the above problem, the sample size leaps to 624!

[1] See p. 88.

Again the auditor can avoid the chore of calculation by looking up a set of tables. Brown and Vance (1961) provide a comprehensive set of tables for this purpose.

## A NOTE ON JOINT CONFIDENCE LIMITS

If we are 95% sure that the proportion of debts which are doubtful lies between, say, 8% and 12%, and that the average value of a doubtful debt lies between, say, £48 and £52 with a 95% level of confidence, we are 95% × 95% = 90% confident that the total value of doubtful debts lies between what two values?

Suppose 8% represent 800 debts and 12% 1200 debts.

It might be thought that by multiplying the upper and the lower bounds of the two estimates together we have our answer, i.e.

$$800 \times £48 = £38{,}400$$
$$1200 \times £52 = £62{,}400$$

But this confidence interval is *much* too wide. If we think for a moment we will see that the chance of 800 and £48 both coming up on the same draw is 20 × 20 = 400. These are 99·5% confidence limits!

The precise calculation of the confidence limits is complicated and beyond the scope of this book. However it almost invariably happens in accounting work that most of the variance is accounted for by the estimate of the proportion rather than the value.

The simplest procedure then is to estimate the value to within a narrow precision limit and then take the midpoint of this limit as a precise estimate of the value. The confidence limits will then be set by the proportion.

If the debts are stored on a random access computer the auditor could run a Monte Carlo simulation to estimate the confidence limits.

Note that the MUS system (Chapter 17) solves this problem by transforming a variable into a binomial distribution.

## SUMMARY

Estimation sampling of variables estimates a value, estimation sampling of attributes estimates a proportion.

The precision limit on an estimate of value or a proportion can be calculated using relatively simple formulae. But since comprehensive sampling tables are readily available the auditor will probably prefer to use these to discover sample size or the precision limit on an estimate.

## QUESTION SERIES 8

1. Given:

| | |
|---|---|
| Population size | 1000 |
| SD of population | £10 |
| Level of confidence | 99% |
| Unit precision limit | £2 |

Calculate required sample size to test if total of inventory is £100,000. (The mean value of your sample comes out to be £103·4. What do you do?) (*)

2. Given:

| | |
|---|---|
| Population size | 2000 |
| SD of population | £10 |
| Level of confidence | 90% |
| Unit precision limit | £0.5 |

Calculate required sample size to test if total of hire purchase debt is £20,000. If you cannot afford to test more than 400 units what can you do? (*)

3. A client states that 10% of his debts are doubtful. You select 500 debtors from a population of 10,000. In 65 cases you estimate the debt to be doubtful. Would you accept the client's estimate of 10%? (*)

4.

| | Estimated standard deviation of population | Acceptable unit precision limit |
|---|---|---|
| (a) | £10 | £2 |
| (b) | £127 | £5 |
| (c) | £2·4 | £0·11 |
| (d) | £365 | £18 |

Calculate required sample sizes to provide 90% level of confidence in the inference. Use formula provided on p. 80, and assume that the population is very large. (*)

5. What does estimation sampling of attributes attempt to do? Give four examples taken from auditing practice.
6. The yearly stock turnover of ten types of inventory is 3·4, 6·1, 8·2, 1·1, 0·7, 6·3, 2·7, 1·8, 7·1, 3·2. What proportion of the population has a turnover under 5? If 1 equals turnover under 5 and 0 equals turnover of 5 or above, what is the attribute mean of the population? (*)

7.

| | Sample size | Error percentage in sample | Level of confidence required (one tailed) | Maximum error rate in population |
|---|---|---|---|---|
| (1) | 200 | 1 | 95 | ? |
| (2) | 300 | 4 | 95 | ? |
| (3) | 250 | 0 | 90 | ? |
| (4) | 100 | 5 | 95 | ? |
| (5) | 500 | 3 | 99 | ? |
| (6) | 350 | 2 | 99 | ? |

Given the above data calculate the maximum likely error rate in the population at the required level of confidence. Use Figures 21.1.1 to 21.1.3. to find maximum error rate. Remember that figures 21.1.1. to 21.1.3. are one tail confidence levels, i.e. they tell us the probability that the population error rate *exceeds* a given %. (*)

8.

| Population size | Maximum acceptable percentage | Level of confidence (one tailed) | Expected percentage | Required sample size |
|---|---|---|---|---|
| 10,000 | 8 | 90 | 4 | ? |
| 15,000 | 2·5 | 90 | 1 | ? |
| 100,000 | 7 | 95 | 3 | ? |
| 50,000 | 11 | 95 | 5 | ? |
| 12,000 | 5 | 99 | 1 | ? |
| 2000 | 28 | 95 | 20 | ? |
| 10,000 | 45 | 97·5 | 40 | ? |

Given the above data use Figures 21.1.1 to 21.2.3 to calculate the sample size needed to provide the required level of confidence. Note that the level of confidence in Figures 21.1.1 to 21.1.3 is one tailed and 21.2.1, to 21.2.3 two tailed. (*)

9. To calculate sample size in estimation sampling of attributes we need to guess the likely percentage in the population having the condition. How can we ensure that this guess is 'conservative'? That is how can we ensure that we do not draw too small a sample? (*)

## SOME ANSWERS TO QUESTION SERIES 8

1. Sample size 150. Recheck estimate of SD. If SD not disproved reject hypothesis that value of inventory is £100,000.
2. Sample size 700. Check skewness. Stratify or widen precision limit.

3. No. The SE on 10% is 1·34. The 95% confidence interval is thus 7·38 to 12·62. A percentage outside these limits will turn up only about 5% of the time in a sample of 500 if 10% of the population of 100,000 is doubtful.
4. (a) 67 (b) 1734 (c) 1285 (d) 1107.
6. 0·6 in both cases.
7. 3%, 6·3%, 1%, 10·3%, 5·3%, 4·7%.
8. 98, 240, 120, 80, 170, 90, 500.
9. Always guess closer to 50% than you believe the actual percentage to be. If you think it will be 2%, say, guess 3%. Do not guess too high or you will make your sample size unreasonably high.

# 9

# Acceptance and Discovery Sampling

### GLOSSARY OF ABBREVIATIONS USED IN ACCEPTANCE SAMPLING

| | |
|---|---|
| MUER | Minimum unacceptable error rate. |
| MURR | Maximum unacceptable rejection rate. |
| A/R | Accept or reject. |

### INTRODUCTION

Acceptance and discovery sampling are two simplified forms of estimation sampling of attributes. They both provide less information than estimation sampling of attributes but, under certain circumstances, they may provide *sufficient* information at an economical sample size.

### ACCEPTANCE SAMPLING

Acceptance sampling is a technique which enables us to accept or reject a *batch* of goods or documents under certain conditions.

Some early advocates of the use of statistical sampling in auditing believed that acceptance sampling was the most useful sampling method for auditors. Vance and Neter (1956), for example, devote no less than one half of their 282 pages to this technique.

Subsequent research has shown this early enthusiasm to be rather misplaced. Acceptance sampling, for reasons which will be explained later in this chapter, is of limited value to the *external* auditor. The method can be of much use to the *internal* auditor.

Acceptance sampling was devised by statisticians to enable storekeepers to control the quality of input to their store. Let us first examine acceptance sampling in this setting and later examine the possibility of transposing the method into an audit context.

The major advantage of acceptance sampling is that it allows all of the analytical work to be done *before* the random sample is drawn. This means that the person who draws the sample and makes an inference from it need have no knowledge of the theory of statistical sampling. He simply draws the sample from the batch and counts the number of defectives in the sample. If the number of defectives exceeds a given figure he rejects the batch, otherwise he accepts it.

The key variable is thus the *maximum* number of defectives the stores inspector can find before he rejects the batch. How do we calculate this key figure?

The problem is best illustrated by an example.

Suppose a stores department receives electrical components from a supplier in batches of 1000. The company wish to control the quality of these components but a 100% check of each batch would be prohibitively expensive. The company admit that a certain proportion of defective components is inevitable but they wish to ensure that in any given batch the percentage of defectives does not exceed 10% of the batch.

If a random sample of 100 units is drawn from the batch how many defectives should the inspector let pass before he rejects the batch?

The answer may seem obvious to the reader. If 10% is the maximum allowable number of defectives the inspector should reject the batch once he locates 11 defectives in the sample!

Unfortunately it is not as easy as this. If the reader turns to Table 3.1 on p. 15 he will see that if the inspector follows such a policy and if every batch has 10% defectives (i.e. all acceptable) the inspector will reject approximately 42% of all batches!

This apparent paradox arises because of sampling error. In 42% of cases the sample proportion of defectives will exceed the population proportion from which it is drawn.

Clearly the inspector cannot reject such a high proportion of 'acceptable' batches.

An alternative strategy might be to choose a higher cut-off rate, say 15%. If the inspector accepts all batches with 15% or less defectives in the sample and all batches have 10% defective, he will only reject about 4% of the batches. This is a big improvement on the previous 42%!

However, although he has gained on the swings he has lost on the roundabouts! By pushing up the acceptable sampling limit to 15% he will accept rather a large number of batches with a defective percentage in *excess* of 10%.

This is the key problem in acceptance sampling and the reason why it is not well suited to audit work.

But before we elaborate this point further let us take a look at the mechanics of acceptance sampling.

## THE MECHANICS OF ACCEPTANCE SAMPLING

The problem in acceptance sampling is to select a suitable sampling plan from among the many sampling plans available. A sampling plan is defined by three variables.

1. The size of the batch, *B*.
2. The size of the sample, *s*.
3. The maximum number of defectives, *d*, which can be found before the batch is rejected.

Suppose cards are being punched in batches of 1000. We wish to select a sampling plan which guarantees that if we select a random sample of *s* units and find *d or less* units defective then we have a 90% level of confidence that the batch contains less than 2% of defective cards.

We use a table such as that set out on pp. 258–260 to select a plan. First we find a table i.e. Table 21.3.2 for the appropriate batch size, in this case 1000. Second we look along the 'error rate in population' column until we find 2%. Third we move down the 2% column until we find a probability of acceptance close to (100 – level of confidence required). In this case the figure we require is (100 – 90) = 10%. The figure of exactly 10% does not occur in the 2% column but several figures close to it do occur, 11·9%, 7·3%, 3·7%, 6·7%, for example. Each of these figures represent a different sample size. 11·9% represents the *smallest* sample size so we will take this plan.

Looking horizontally along the row containing 11·9% we find the figures '100,0', The 100 represents *s*, the sample size, the 0 represents, *d*, the maximum acceptable number of defectives.

Thus our sampling plan is (1000, 100, 0). This means that if we select a sample of 100 from a batch of 1000 and find no defectives we can be (100 – 11·9) = 88·1% confident that the error rate is less than 2%.

If we had selected instead the sampling plan (1000, 200, 1), we would have ensured that if we found 1 or 0 defectives in the random sample of 200 then we could be (100 – 6·7%) = 93·3% confident that the error rate was less than 2%.

Since a doubling of sample size only increases our level of confidence from 88·1% to 93·3% we would be likely to adopt the plan (1000, 100, 0).

The reader may care to select suitable acceptance sampling plans for a batch of 1000 under the following conditions.

| Minimum unacceptable error rate | Level of confidence | Suggested plan |
|---|---|---|
| 1% | 95% | (1000, 275, 0) |
| 3% | 90% | (1000, 75, 0) |
| 10% | 99% | (1000, 50, 0) |
| 2% | 99% | (1000, 200, 0) |

In every case the most suitable sampling plan has $d = 0$. We shall see in a moment why this is so, and why it would *not* be so in practice!

The reader will have noticed that all of the sampling plans discussed above guarantee that we will reject *unacceptable* batches with a given level of confidence. But what about rejecting *acceptable* batches? A sampling plan that always rejected all batches would clearly reject all unacceptable batches. But I hope the reader will agree that it would be of rather limited use!

Let us illustrate the problem with an analogy.

## THE CASE OF THE 'HARDLINE' POLICEMAN

Suppose that an international rugger match is being held and one policeman is placed at either of the two entrances to the rugby ground. The policemen are told that demonstrations are likely to take place and so all likely agitators should be excluded from the stadium.

One policeman adopts a 'hardline' policy of excluding all persons carrying placards, wearing beards or Afro hairstyles, disseminating a ripe odour and so on.

The other policeman who holds liberal views only excludes persons carrying inflamatory placards.

1000 spectators pass by each policeman and the result of their respective vetting policies are given in Table 9.1.

Which policeman did the better job? The reader may care to spend a little time on Table 9.1 trying to answer this question.

A careful study of Table 9.1 will show that the question cannot be answered unless the word 'best' is carefully defined.

The hardline policeman has put up the better performance in keeping demonstrators out but at the cost of excluding 102 innocent spectators.

The liberal policeman has excluded no innocent spectators but has failed to detect 50 demonstrators.

This is exactly the problem that faces the acceptance sampler. It is very easy to find an acceptance sampling plan which will *either* pick up most batches with an error rate exceeding $X\%$, *or* which will *not* reject batches with an error rate less than $(X - Y)\%$. The problem is that in audit work it is difficult to find a sampling plan which will achieve both objectives at one and the same time.

Table 9.1 Results of vetting procedure of two policemen.

| | Policeman | |
|---|---|---|
| Spectators | Liberal | Hardliner |
| Allowed in: | | |
| Non-demonstrators | 900 | 798 |
| demonstrators | 50 | 2 |
| Excluded: | | |
| Non-demonstrators | 0 | 102 |
| demonstrators | 50 | 98 |
| | 1000 | 1000 |

Let us take a specific example.

Priced vouchers batched by the 1000 are processed by a pricing department. The external auditor wishes to be 90% sure that the error rate in pricing does not exceed 3% of the items priced. On the other hand he wants to be 95% sure that he will not reject a batch with a pricing error rate of 1% or below.

He consults a set of acceptance sampling tables,[1] such as those on p. 259, and looks up the table for:

| | |
|---|---|
| Population size | 1000 |
| Minimum unacceptable error rate | 3% (90% confidence) |
| Maximum unacceptable rejection rate | 1% (95% confidence) |

The sampling plan (1000, 75, 0) gives the auditor a 90·7% level of confidence of rejecting all batches with an error rate of 3% or more. But it only provides a 45·7% level of confidence of *accepting* batches with a 1% error rate as shown in Figure 9.1. This is likely to be unacceptable. Too many good batches would be rejected.

We must increase our sample size to (1000, 275, 5) to achieve a confidence level of 96·9% in accepting batches with the maximum unacceptable rejection rate of 1%. This certainly satisfies our minimum unacceptable error rate, giving us, in fact, a 94·2% level of confidence, 4·2% higher than required and so rather wasteful of auditing resources.

[1] Or he devises his own plan using the table constructed by J. M. Cameron which is set out in Table 21.4 on page 261.

The reader may care to try to find a suitable acceptance sampling plan for:

| | |
|---|---|
| Population | 1000 |
| MUER | 2% (90% level of confidence) |
| MURR | 1% (95% level of confidence) |

(1000, 300, 3) satisfies the MUER but only provides a 65% level of confidence in the MURR.

As the MUER approaches the MURR the sample size grows larger. When they are equal the sample size is equal to the batch size!

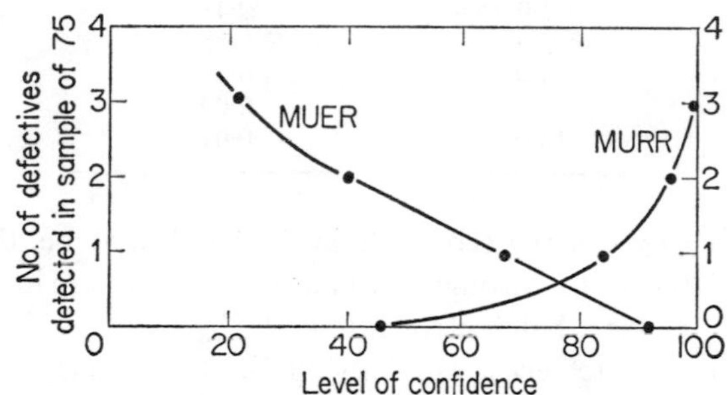


Fig. 9.1. The graph shows the level of confidence of rejecting batches with 3% error or above (MUER curve) along with the level of confidence of not rejecting batches with error rates as low as 1% (MURR curve).

| | |
|---|---|
| Population | 1000 |
| MUER | 3% |
| MURR | 1% |
| Sample size | 75 |

Notice that the levels of confidence of the two curves run counter to one another.

Several solutions have been put forward to solve this problem.

One approach is to carry out a 100% check on all rejected batches. In this way the auditor can eliminate the clients' risk by picking up all acceptable batches which were rejected in error. Thus the only errors remaining in the batch are the errors in the acceptable batches. If we suppose that *all* the incoming batches actually contain 3% of error the number of undetected errors in the long run for the (1000, 75, 0) plan would be

$$\frac{(0{\cdot}907 \times 0) + (0{\cdot}093 \times 0{\cdot}03)}{0{\cdot}907 + 0{\cdot}093} = 0{\cdot}0028.$$

This means that 0·28% of errors will slip through the system in the long run *if all rejected batches are 100% checked.*

However the percentage of error in the batches will not always be 3%. The proportion of error that will slip through the control system for various batch error rates is shown below.

| Batch error rate | Percentage of error which will slip through |
|---|---|
| 0·1 | 0·09 |
| 0·5 | 0·34 |
| 1·0 | 0·45 |
| 1·5 | 0·46 |
| 2·0 | 0·41 |
| 3·0 | 0·28 |
| 4·0 | 0·16 |
| 5·0 | 0·09 |
| 10·0 | 0·01 |

Notice that *whatever* the error rate in the batch may be, the maximum error rate which can slip through if all rejected batches are inspected 100% is around 0·46% for a batch error rate of 1·5%.

This maximum batch error rate only applies to the acceptance sampling scheme (1000, 75, 0), but an auditor can calculate the maximum batch error rate for any other sampling plan in a few minutes by using a set of tables like Tables 21.3.1 to 21.3.3 on pp. 258–260. Arkin (1963) provides a comprehensive set of tables for the purpose.

In summary, by carrying out a 100% check on all rejected batches the auditor can limit the maximum average error rate in the audited batches to a given figure. The actual error rate allowed through is likely to be a good deal lower than this maximum possible error rate.

## THE DOUBLE SAMPLING APPROACH

Another method of improving the efficiency of acceptance sampling is to to take two bites at the cherry. This method is particularly appropriate where there is great variety between the quality of the batches being audited.

Let us take an example.

Suppose we are auditing several batches of 1000 documents and the audit requirements are as follows.

| | |
|---|---|
| Population size | 2000 |
| MUER | 3% (90% CL) |
| MURR | 1% (90% CL) |

A plan of (2000, 300, 5) is needed to satisfy this audit scheme. If however many of the batches are either of very good quality, say less than 0·5% error, or of very bad quality, say more than 10% error, an acceptance sample of very much smaller size could differentiate between them, i.e.

| | |
|---|---|
| Population | 2000 |
| MUER | 10% (90% CL) |
| MURR | 0·5% (90% CL) |

A plan of (2000, 50, 1) is sufficient to satisfy these requirements. The required sample size has been cut by 80%!

In double acceptance sampling the auditor first applies a sampling plan of lower discrimination to the batch. If it is *rejected* by this plan it is finally rejected. If it is accepted it must then pass the second tougher acceptance sampling plan to be finally accepted.

By using double sampling[1] the auditor avoids wasting his time checking batches with very high error rates.

In the example quoted he would check a random sample of 50 units. If more than one error is found he will reject the batch. If one error or less is found he will draw another 250 units and reject if more than 5 errors are found in total.

## SEQUENTIAL SAMPLING

If double sampling can improve the efficiency of acceptance sampling then why not have triple, quadruple, or quintuple sampling?

This is possible, although it may present certain operational problems if there is an unavoidable delay between drawing a sample and checking it.

We can leap straight to the logical conclusion of this process by using *sequential sampling*.

Sequential sampling can best be explained by using a diagram such as Figure 9.2. The auditor begins to draw a random sample. So long as the acceptance–rejection line falls in the white area he continues to sample. Once it crosses either line he either accepts or rejects the batch.

In the example given the 40th and 70th units audited contain an error but after 150 units have been audited sufficient correct units have been found to accept the batch.

The construction of sequential sampling graphs is discussed in Vance and Neter (1956), pages 109–116.

If, after a certain maximum sample size has been reached, the accept/

[1] A table for devising double sampling plans constructed by Paul Peach is provided in Table 21.5 on page 265.

reject line is still in the white area, the auditor should accept if the line is closer to the accept line and vice-versa.

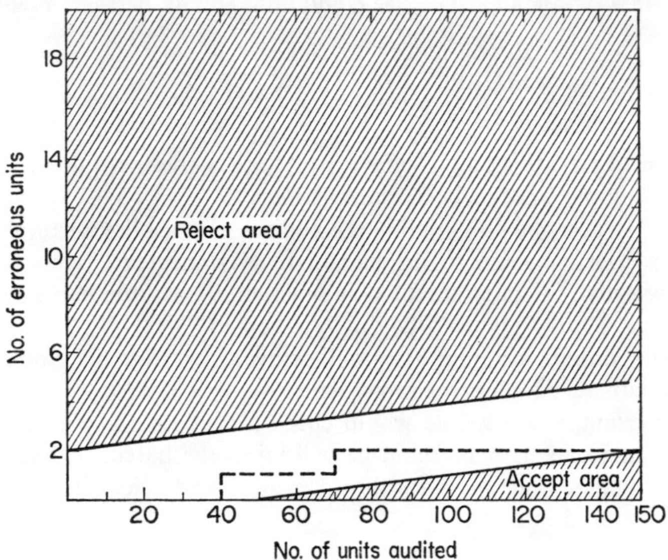


Fig. 9.2. Sequential sampling acceptance table. The 40th and 70th units contain an error but the batch is eventually accepted after 150 units have been audited.

## WHICH ACCEPTANCE SAMPLING PLAN IS THE BEST?

We have discussed single, double and sequential sampling. Which is the better method for an auditor to use?

If by better we mean most efficient in the sense of minimizing average sample size then sequential sampling is clearly the most efficient whatever the error rate in the population.

The statistical research group at Columbia University has calculated that where the condition sought for, i.e. error makes up a low proportion of the population, say under 5%, then double sampling is about 25% more efficient than single sampling and sequential sampling is about 40% more efficient.

The writer has found sequential sampling rather awkward to use in auditing since the auditor must pull an initial small sample, 50 in the example given, and then pull random samples *one at a time*, until the A/R decision is taken.

Double sampling is simpler to use and I recommend it.

The double sampling plan can be selected as if two single plans were being selected. Obviously the acceptance number in the second plan must exceed the acceptance number in the first. As an alternative one can use Peach's table set out in Table 21.5 on p. 265.

## DISCOVERY SAMPLING

Discovery sampling is normally used by an auditor for carrying out a *preliminary* test on the quality of a series of batches or populations of data. For this reason the method is sometimes called *exploratory* sampling.

The reader will recall that in the series of examples of acceptance sampling that we examined on p. 97 the best acceptance sampling plan *in every case* had a defective number of zero.

This was no coincidence. At this point in our discussion we were concerned only with the MUER. We wanted to ensure that we rejected all batches with an unacceptable error rate with a given level of confidence.

Once we introduced the concept of the MURR the situation changed and the acceptable number of defectives in the sampling plan varied from problem to problem.

Discovery sampling is only concerned with the MUER so the number of acceptable defectives in a discovery sampling plan is always zero.

The reader could build up a discovery sampling table by selecting out all of the sampling plans with $d = 0$ from the acceptance sampling tables on pp. 258–260.

But how do we use discovery sampling as a screening device? Again we will attempt to illustrate the principle by using an analogy.

## THE CASE OF THE OVERWORKED DOCTOR

Once upon a time, in a remote part of the South Seas, there lived a very overworked doctor.

The trouble was that a disease called Arkin's disease had broken out in the islands. It is a peculiarity of this disease that if 2% or more of the population catch the disease it quickly assumes epidemic proportions. If less than 2% of the population have the disease it is unfortunate for these few individuals who suffer from it, since it is nearly always fatal, however it is unlikely to spread to the rest of the population.

There were five islands under the doctor's administration and the population of these islands varied from a few hundred to several thousand.

A vaccine against the disease was available but it was expensive.

The doctor determined to vaccinate only those islands where it seems that 2% or more of the population have the disease. But how can the doctor determine whether or not the proportion of the population with the disease is equal to or in excess of the critical 2%?

The doctor is familiar with scientific sampling methods and he knows that an estimation sampling of attributes to test for a 2% proportion will require rather a large sample.

For example:

| | |
|---|---|
| Population | 2000 |
| Expected percentage having disease | 2% |
| Precision limit | ±1% |
| Confidence level | 99% |

would require a random sample of 790 units!

The doctor's problem is that he requires a high level of confidence in estimating a very small proportion. A situation not unfamiliar to the auditor.

By good fortune, however, the doctor has heard of the technique of discovery sampling. So he consults a suitable set of discovery sampling tables, left to him by an unfortunate accountant who had expired from Arkin's disease in the previous year, and finds that in the situation

| | |
|---|---|
| Population | 2000 |
| Minimum unacceptable error rate | 2% |
| Level of confidence | 99% |

the requisite sample size is only 210 for a level of confidence of 99%.

The other islands have populations of 500, 1000, 5000, 10,000 and 100,000.

The sample size in these islands under the same conditions along with the sample size using estimation sampling of attributes is

| | | Sample size | |
|---|---|---|---|
| Island | Population | Discovery sampling | Est. sampling of attributes |
| 1 | 500 | 197 | 260 |
| 2 | 1000 | 204 | 536 |
| 3 | 2000 | 210 | 790 |
| 4 | 5000 | 214 | 1030 |
| 5 | 10,000 | 220 | 1150 |
| 6 | 100,000 | 235 | 1280 |

The doctor selects the required number of random individuals from each island, and checks each individual for the disease. If he finds *one* person with the disease he vaccinates the entire population immediately, i.e. if the first person he checks has the disease he need not check the remaining 199! He goes right ahead and vaccinates. If none of the random sample have the disease he does not vaccinate.

We see that discovery sampling provides the doctor with sufficient information to answer his question at a very economical sample size.

However the maxim 'nothing is for nothing' applies as much to sampling as it does elsewhere. An unrestricted random sample of 200 must provide less information than an unrestricted random sample of 790.

The perceptive reader will have noticed that the discovery sample used above on the population of 1000 is actually the acceptance sampling plan (1000, 200, 0). If we turn to the acceptance sampling tables on p. 259 we see that this plan provides the following probability of acceptance.

| Percentage error | Probability of acceptance |
|---|---|
| 0·5 | 32·7 |
| 1·0 | 10·6 |
| 2·0 | 1·1 |
| 3·0 | 0·1 |

The sampler has provided himself with a (100 − 1·1)% = 98·9% level of confidence that he will reject all populations with an error rate of 2% or above. But if the error rate is as low as 0·5% he will only have a confidence of 32·7% in *accepting* the population.

This is why, as we stated in the opening paragraph, discovery sampling is best used as an economical screening device, a preliminary rough test to pick out batches or populations requiring further investigation.

## PRACTICAL EXAMPLE OF DISCOVERY SAMPLING

Let us work through a practical example.

Lists of inventory are sent in from several branches at the end of the year for stocktaking. The lists consist of quantity, unit value, and the extension to total value.

The auditor cannot afford to check the pricing and extensions of every branch. He decides to use discovery sampling to obtain a 90% level of confidence that the error rate on both pricing and extensions is less than 1%.

There are 2000 items of inventory so the sample size needed to provide a 90% level of confidence is approximately 220.

The auditor draws 220 random extensions, probably using a systematic random sample. If he finds no errors he need not check the remainder of the list. But *immediately he finds one error* he stops sampling and checks all of the extensions on the list.

We repeat that this procedure will entail him checking a good number of lists with error proportions less than 1%. But where the clerical work is very accurate the discovery sample will 'pass' the list.

An alternative scheme is to switch to an alternative form of sampling once discovery sampling indicates that the population may not be acceptable.

For example if the auditor finds an arithmetic error when checking the extension list he can switch to estimation sampling of attribute and calculate the sample size needed to ensure that the population error rate is 1% with a precision limit of say ±1% at a 99% level of confidence. The sample size under these conditions is 800 units.

An alternative plan might be to switch to acceptance sampling by introducing a minimum unacceptable rejection rate. The reader may care to use Cameron's table on p. 262 to devise a suitable plan at MURR of 0·5%.

## SUMMARY

Discovery sampling is a screening device for picking out populations requiring further investigation.

The auditor discovers the population size and decides on (a) the minimum unacceptable error rate and (b) the minimum level of confidence acceptable under the circumstances.

He consults a set of discovery sampling tables and discovers the required sample size.

He draws a random sample of this size and if *no* errors are discovered he 'passes' the population. If an error is located he *immediately* stops discovery sampling and either checks the entire population or switches to another type of estimation sampling.

Discovery sampling provides sufficient information to answer the auditors' questions under certain conditions. We should not, however, overlook the fact that the method will tend to reject a good number of acceptable batches.

Several advocates of discovery sampling have not laid sufficient stress on this last point.

## QUESTION SERIES 9

1. Is acceptance sampling of more use to the internal or external auditor?
2. What is the main operational advantage of acceptance sampling?
3. Use Table 3.1 on p. 15 to calculate the number of good batches a stores inspector would reject if he chose a cut-off point of 13% or above and all batches had 10% of defectives. Batch size = 1000, sample size = 100. (*)
4. Select a suitable acceptance sampling plan for rejecting unacceptable batches under the following conditions. Batch size = 500.

| Minimum unacceptable error rate | Level of confidence required |
|---|---|
| 3% | 99·9% |
| 2% | 95% |
| 1% | 90% |

5. Why is acceptance sampling of limited use in audit work?
6. Describe one application where acceptance sampling is of value to the internal auditor.
7. Devise suitable sampling plans under the following conditions. Use tables on pp. 258–260.

| Batch size | MUER | Level of confidence | MURR | Level of confidence |
|---|---|---|---|---|
| 500 | 10 | 90 | 2 | 95 |
| 500 | 10 | 99 | 1 | 90 |
| 500 | 3 | 95 | 0·4 | 90 |
| 1000 | 10 | 99 | 1 | 90 |
| 1000 | 3 | 90 | 0·5 | 90 |
| 1000 | 2 | 90 | 0·5 | 95 |

(*)

8. What is the normal use to which discovery sampling is put by auditors?
9. A discovery sample of 100 is checked and no errors are found. What does this tell us about the batch? (*)
10. What is the relationship between discovery sampling and acceptance sampling?
11. What is the main disadvantage of discovery sampling? (*)

12. If one error is found in a discovery sample what does the auditor do next?
13. Use Table 21.2 on p. 239 to calculate required discovery sampling sizes under the following conditions.

| | Population size | MUER | Level of confidence |
|---|---|---|---|
| (a) | 500 | 0·5% | 99% |
| (b) | 500 | 2% | 90% |
| (c) | 1000 | 0·5% | 95% |
| (d) | 1000 | 1% | 90% |
| (e) | 10,000 | 2% | 99% |
| (f) | 1,000,000 | 1% | 95% |

(*)

## SOME ANSWERS TO QUESTION SERIES 9

3. 19·74%.
4. With acceptance number of 0.
   (1) 175 (2) 125 (3) 175.
7. (500, 50, 2), (500, 60, 1), (500, 150, 1), (1000, 60, 1), (1000, 125, 1), (1000, 300, 3).
9. We are $a$% confident that the error rate in the batch is less than $x$%.
11. A good number of acceptable batches are rejected.
13. 611, 103, 481, 205, 220, 325.

# 10

# Cluster, Multistage and Replicated Sampling

## THE CLUSTER SAMPLE

With pure and systematic random sampling we select *individual* units from the population. *Cluster sampling* is a method of sampling by which we select *groups* of units, the *first* unit of each group being selected by means of random number tables.

Suppose, for example, that trucking vouchers for several years past have been stored in batches of twenty. A sample of 1000 from a population of 500,000 is required to make an estimate of suspected overcharging during the previous years. Rather than draw 1000 individual trucking vouchers at random from the population, which would be a time-consuming business, we might decide to use cluster sampling and draw 50 sample batches containing 20 units each. We select 50 random numbers and use these to sample from among the 25,000 batches.

Wherever a group of 'backing' documents are attached as proof to a prime document, such as a group of vouchers proving a cheque, cluster sampling is likely to prove the cheaper form of sampling.

Note that we use cluster sampling to cut the cost of sampling and the cost of replacing the sample.

The reader, probably, will have already realized that traditional audit procedures use cluster sampling. The conventional approach to auditing is for the auditor to select, say, a single month of invoices to test the accuracy of invoicing, or a single week's payroll to test the accuracy of the pay-out procedure.

It does not, however, follow from this that an auditor switching to statistical from traditional sampling can continue to use cluster samples. A cluster sample is likely to prove less efficient than an unrestricted random sample of equal size. By less efficient we mean that the inference from the cluster sample is likely to be less accurate.

The reason for this is that the variability within the cluster is likely to be less than the variability in the population as a whole. In other words payslips, sales slips or invoices, etc., contiguous to one another are likely to be similar in some respect. Employees in a given department are likely to have salaries more similar to one another than to employees in the firm as a whole.

The relative variability is a question of fact, which can be tested. If the variability within clusters is no different to the variability of the population *or if it is greater*, then cluster sampling is as efficient or even more efficient and also cheaper to employ, than unrestricted random sampling.

It is however, often found that cluster sampling is so inefficient relative to unrestricted random sampling that in order to achieve the same degree of confidence in the inference the auditor must select a very much larger sample. So much larger, in fact, that it cancels out the relative cost advantage of cluster sampling.

## HOW DO WE CALCULATE THE PRECISION LIMIT ON A CLUSTER SAMPLE?—VARIABLE SAMPLING

In cluster sampling as with the other methods of sampling previously described we must calculate a mean for the individual sample unit and multiply this by the number of units in the population to arrive at an estimate of the value of the population. Secondly we must calculate the precision limit on this estimate.
The method of doing this is as follows.

1. Calculate the value of the average unit in each cluster.
2. Calculate the average of the cluster averages.
3. Multiply the figure arrived at in (2) by the number of units in the population. This provides a best estimate of the value of the population.

But the cluster average like any other sample is subject to sampling error. How do we measure the precision of our estimate? We proceed as follows:

1. Calculate the standard deviation of our cluster averages.
2. Multiply this figure by a factor selected from Table 10.1. The result is a measure of the precision of the estimate.

Let us take as an example the problem of sampling from 25,000 batches of transport vouchers mentioned previously.

The reader will recall that an unrestricted random sample required 1000 units so we decided to sample 50 batches of 20 units each.

Suppose that the error of the average unit of the fifty batches turned out to be +£0·05, so that over the population of 500,000 vouchers our best estimate of total value of error is £0·05 × 500,000 = £25,000. What are the precision limits on this estimate at a 99% level of confidence?

First we must calculate the standard deviation of the various cluster averages. There are 50 cluster averages so we use one of the methods described in Chapter 5 to calculate the standard deviation. This turns out to be £0·03.

Table 10.1. Table to calculate precision limits on cluster sample. The factor given inside the box is the precision limit divided by the standard deviation. (With small samples, say under 50 clusters, we must use Student's *t* distribution rather than the figures derived from the Normal distribution.)

| Number of clusters | Required level of confidence | | |
|---|---|---|---|
| | 99% | 95% | 90% |
| 10 | 1·083 | 0·753 | 0·610 |
| 20 | 0·656 | 0·480 | 0·397 |
| 30 | 0·512 | 0·379 | 0·316 |
| 40 | 0·434 | 0·324 | 0·271 |
| 50 | 0·381 | 0·286 | 0·238 |
| 60 | 0·345 | 0·261 | 0·214 |

Next we look up Table 10.1 and find that for 50 clusters at a 99% level of confidence the relevant factor is 0·381. We multiply the standard deviation by this factor.

$$£0{\cdot}03 \times 0{\cdot}381 = 0{\cdot}01143$$

This gives us the confidence interval of the unit average at a 99% level of confidence. To calculate the confidence interval at a 99% level of confidence for the population as a whole we multiply this figure by 500,000, i.e.

$$500{,}000 \times (0{\cdot}03 \times 0{\cdot}381) = £5715 \qquad \text{(A)}$$

So we can be 99% sure that the actual value of error lies between £25,000 ± £5715 that is between £19,285 and £30,715. But suppose we found this degree of reliability unacceptable. What could we do?

The simplest procedure would be to substitute the maximum acceptable confidence limit into formula (A) above and substitute a symbol, say *f*, for the factor from Table 10.1. Suppose the maximum acceptable precision limit is ± £4000, then,

$$500{,}000 \times (0{\cdot}03 \times f) = £4000$$
$$f = 0{\cdot}266$$

Returning to Table 10.1 we find that a factor of 0·266 gives us close to a 95% level of confidence with 60 clusters or a 90% level of confidence with about 43 clusters.

By playing around with various permutations we should be able to arrive at an acceptable sampling plan.

When the number of clusters exceeds fifty we can use large sample tables, that is the conventional tables we used for calculating sample size in variables sampling as set out on Figures 21.5.1 to 21.5.8.

For example with

| | |
|---|---|
| Population | 20,000 clusters |
| Standard deviation | 0·03 |
| Precision limit (unit) | 0·008 (£4000 ÷ 500,000) |

the required number of clusters using large sample tables is:

| Confidence level | No. of clusters |
|---|---|
| 99% | 105 |
| 95% | 61 |
| 90% | 43 |

## CLUSTER SAMPLING FOR ATTRIBUTES

The previous section discussed the use of cluster sampling to measure a variable such as the value of inventory or error. Cluster sampling can also be used to measure a proportion, i.e. the proportion of a population with a given attribute.

Applying cluster sampling to measuring attributes presents no additional problems.

The procedure is as follows.

1. Calculate the percentage of each cluster which has the given condition.
2. Calculate the average of the cluster averages.
3. Calculate the standard deviation of the cluster averages.
4. Multiply the standard deviation by factor from Table 10.1 to calculate the unit precision limit as before.

5. Multiply unit precision limit by number of units in the population to arrive at precision limit for estimate of population proportion.

To take an example,

| | |
|---|---|
| Population | 10,000 |
| No. of clusters | 30 × 10 units |
| Confidence level | 95% |

We carry out steps 1, 2 and 3 above and find that the average of the sample averages is 20% and the standard deviation 5%.

From Table 10.1 we select factor 0·379, i.e. 30 clusters at a 95% level of confidence. We multiply the standard deviation by this factor.

$$5\% \times 0{\cdot}379 = 1{\cdot}895$$

which gives us a precision limit of almost ±1·9% on 10%. In other words we are 95% confident that the actual proportion of the population having the condition lies within the limits 8·1% to 11·9%.

## SUMMARY

In summary, auditors traditionally use cluster sampling but the number of clusters they draw is usually very much less than the *minimum* number of clusters needed in statistical sampling. *At least* twenty clusters will be needed.

Cluster samples simplify the operational problem of drawing and replacing the sample but since the variability within the cluster is likely to be less than the variability within the population, cluster sampling is likely to provide a less accurate inference than unrestricted random sampling of the same sample size.

## MULTISTAGE SAMPLING

Another special type of sampling technique is called *multistage* sampling. Figure 10.1 illustrates the type of situation where this form of sampling can be useful.

Suppose that a company operates six branches which are widely dispersed geographically. Each branch employs a number of salesmen varying from five to eleven as illustrated in Figure 10.1.

The company decides to alter the existing scheme for paying sales commission, and higher management want to collect a cross-section of salesman opinion on the new scheme. They decide to interview a dozen salesmen on the topic.

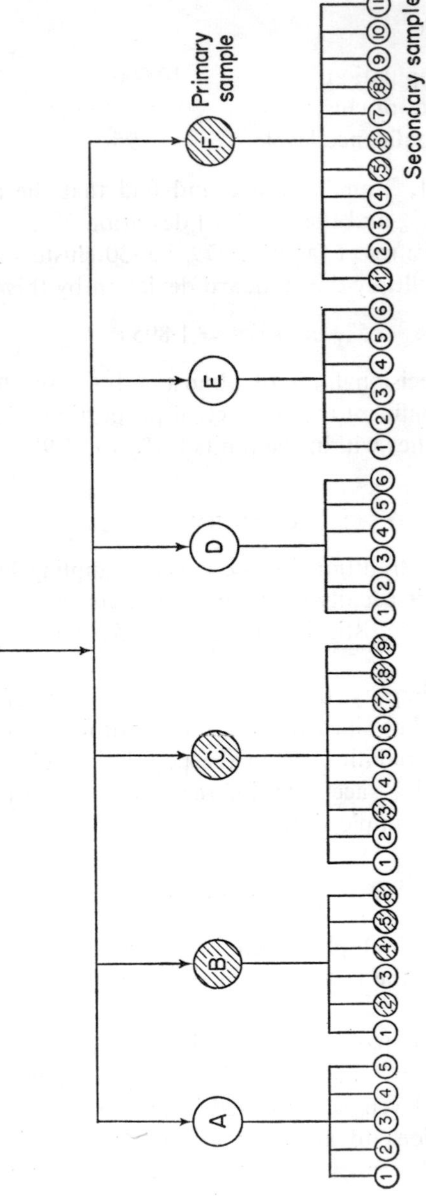


Fig. 10.1. Diagram of a multistage sample of 12 units from population of 43 units. The units sampled are shaded.

It would be a simple matter to allocate numbers 1–43 to the forty-three salesmen and draw twelve numbers between 1 and 43 from random number tables to select a random sample. However, this might entail visiting all six branches, an expensive and time consuming business since the branches are geographically dispersed over a wide area.

Instead the company sampler decides to draw a multistage sample. At the first stage he draws a random sample of three of the six branches, at the second stage he draws a random sample of four salesmen from each branch selected at the first stage.

In the example given branches B, C and F are selected and within each branch the following salesmen are selected:

| Branch | Salesmen |
|---|---|
| B | 2, 4, 5, 6 |
| C | 3, 7, 8, 9 |
| F | 1, 5, 6, 8 |

Notice that in the example given we have no reason to suspect that the opinion of the salesmen in one branch will differ significantly from the opinion of salesmen in any other branch. So far as we know the population of salesmen is homogeneous as far as their opinion of the new commission scheme is concerned. If we have reason to believe that the opinion of salesmen in one branch will differ significantly from the opinion of salesmen in other branches we should *stratify* our population by selecting the odd man out for special investigation.

We should also notice that by employing a two stage sample we have introduced the possibility of sampling error at *two* points in the sampling programme.

Let us now examine another example which is formally identical to the first.

A retail chain of stores run 100 branches and maintain stock records at head office. The internal auditor wishes to test the reliability of the total value of the head office inventory record by physically checking 300 lots of inventory from a total of 100,000 stocked by the 100 branches. He intends to carry out the check over one weekend.

An unrestricted random sample is out of the question since it would entail so much travelling. The internal auditor decides to select a two-stage sample.

At the first stage he will select 10 branches at random. At the second stage he will select 30 random lots for checking from each of these 10 branches.

The data from this test is set out in Tables 10.2 and 10.3.

First the auditor must calculate the mean value of the lot in each group of 30 sampled. This is given in column (b) of Table 10.2. The average value of the averages of the 30 lots is £150.

Table 10.2. Data on multistage sample (1).

| (a) Store | (b) Mean value of 30 lots | (c) Difference from mean | (d) (Difference)$^2$ |
|---|---|---|---|
| 1 | 157 | +7 | 49 |
| 2 | 150 | — | — |
| 3 | 173 | +23 | 529 |
| 4 | 121 | −29 | 841 |
| 5 | 154 | +4 | 16 |
| 6 | 161 | +11 | 121 |
| 7 | 132 | −18 | 324 |
| 8 | 148 | −2 | 4 |
| 9 | 173 | +23 | 529 |
| 10 | 131 | −19 | 361 |
| Mean or total | 150 | 0 | 2774 |

Thus:

$$\text{standard deviation} = \left(\frac{2774}{10}\right)^{1/2} = £16{\cdot}66^{a}$$

$$\text{sampling error} = 16{\cdot}66 \times 0{\cdot}610 = £10{\cdot}163$$

[a] When SD is computed for entire population we divide by $n$, if from sample by $(n - 1)$.

Since there are 100,000 lots altogether the best estimate of the value of all of the lots is 100,000 × £150 = £15,000,000. But how accurate is this estimate? What are the precision limits at a given level of confidence?

This presents us with a problem.

The trouble is that we have introduced sampling error at *two* points in the sampling plan. The first stage sample of the branches introduced some sampling error. The second stage sample of lots within each branch introduced more sampling error.

To calculate the sampling reliability of the estimate we must add these two amounts of sampling error together.

First let us calculate the sampling error of the averages of the ten branches selected. This is done in Table 10.2 above. The standard deviation of 10 lot averages turns out to be £16·66. The factor to adjust for sample reliability from Table 10.1 is 0·610 at a 90% level of confidence. The sampling error is thus:

$$16{\cdot}66 \times 0{\cdot}610 = £10{\cdot}163$$

Turning now to the second part of the problem we need to calculate the standard deviation of each sample of thirty for each store. The auditor makes this calculation and the results are provided in column (e) Table 10.3. We now multiply these standard deviations by the sample reliability factor from Table 10.1. The relevant factor is 0·316 for a sample of 30 at a 90% level of confidence.

Table 10.3. Data on multistage sample (2).

| (a) Store | (b) No. of lots stocked | (c) Size of sample | (d) Factor from Table 10.1 | (e) Standard deviation of sample | (f) Column (d) × column (e) | (g) Column (f) squared |
|---|---|---|---|---|---|---|
| 1 | 936 | 30 | 0·316 | 43 | 13·59 | 184·69 |
| 2 | 721 | 30 | 0·316 | 28 | 8·85 | 78·32 |
| 3 | 1227 | 30 | 0·316 | 35 | 11·06 | 122·32 |
| 4 | 846 | 30 | 0·316 | 36 | 11·38 | 129·51 |
| 5 | 1111 | 30 | 0·316 | 49 | 15·48 | 239·63 |
| 6 | 894 | 30 | 0·316 | 41 | 12·96 | 167·96 |
| 7 | 1206 | 30 | 0·316 | 18 | 5·69 | 32·38 |
| 8 | 997 | 30 | 0·316 | 36 | 11·38 | 129·51 |
| 9 | 1054 | 30 | 0·316 | 44 | 13·91 | 193·49 |
| 10 | 1008 | 30 | 0·316 | 31 | 9·80 | 96·04 |
| Mean or total | 10,000 | 300 | | 36·1 | | 137·39 |

We have now reached a stage where we can examine the formula for calculating the sampling error of the estimate of a multistage sample. The required formula is:

$$S = \left(r^2 + \frac{n}{N}a\right)^{1/2}$$

when $S$ = sampling error of mean estimate of total, $r$ = sampling error of the lot average of the ten branch averages, $a$ = average of the squares of the individual sampling errors of the ten branches, $n$ = number of branches sampled, $N$ = total number of branches which could have been sampled.

Substituting the figures calculated in Tables 10.2 and 10.3

$$S = \left((10{\cdot}163)^2 + \frac{10}{100}(137{\cdot}39)\right)^{1/2}$$

$$= (103{\cdot}29 + 13{\cdot}74)^{1/2}$$

$$S = (117{\cdot}03)^{1/2} = £10{\cdot}82$$

Since there are 100,000 units in the population the sampling error of the total is:

$$£10{\cdot}82 \times 100{,}000 = £1{,}082{,}000$$

Therefore we can state with a 90% level of confidence that the value of inventory lies between £15,000,000 ± £1,082,000, or roughly between £14m and £16m.

Suppose we had not taken a multistage sample but simply selected 300 random lots from a population of 100,000 lots with a mean value of £150 and a standard deviation of £36·1? What would our confidence interval have been at a 90% level of confidence?

The answer turns out to be £15,000,000 ± £350,000 approximately. The multistage sample has saved travelling time but at the cost of reducing the precision of the inference.

## A GENERAL METHOD OF CALCULATING SAMPLING ERROR

We have emphasized several times in this book that the major problem in sampling is to calculate the sampling error on the best estimate of total value or proportion.

In the previous section we discovered that the calculation of a confidence interval on an estimate based on a two-stage sample was quite complicated. The calculation based on a three-stage sample, which would be rare in auditing, is even more complicated. Further, it may happen that we wish to combine various sampling strategies such as a multistage cluster sample. In this case the calculation of sampling error would become even more complicated.

The reader will recall that a rather similar situation arose when we discussed the calculation of the standard deviation. Various precise methods of calculation were available when the population was relatively small, but when the population was very large we switched to the relatively simple 'average range method' of *estimating* the standard deviation.

When a rather complicated sampling plan is in operation it may be easier to use a method called *replicated sampling* to calculate the sampling error of the estimate rather than use the methods previously described.

The reader will note that no matter how complex the sampling plan

Table 10.4. Data on replicated sampling problem.

| Group | (a) Mean value of unit in each group | (b) Deviation from mean | (c) (Deviation)$^2$ |
|---|---|---|---|
| 1 | 136 | −14 | 196 |
| 2 | 180 | +30 | 900 |
| 3 | 169 | +19 | 361 |
| 4 | 119 | −31 | 961 |
| 5 | 173 | +23 | 529 |
| 6 | 131 | −19 | 361 |
| 7 | 175 | +25 | 625 |
| 8 | 154 | +4 | 16 |
| 9 | 155 | +5 | 25 |
| 10 | 160 | +10 | 100 |
| 11 | 140 | −10 | 100 |
| 12 | 173 | +23 | 529 |
| 13 | 164 | +14 | 196 |
| 14 | 143 | −7 | 49 |
| 15 | 129 | −21 | 441 |
| 16 | 140 | −10 | 100 |
| 17 | 116 | −34 | 1156 |
| 18 | 145 | −5 | 25 |
| 19 | 152 | +2 | 4 |
| 20 | 148 | −2 | 4 |
| | 150 | 0 | 6678 |

Mean = £150

$$\text{Standard deviation} = \left(\frac{6678}{20}\right)^{1/2} = £18{\cdot}27$$

Sampling error = £18·27 × 0·397 = £7·25
(at 90% confidence level)

may be the final result is simply $n$ units sampled from a population of $N$ units. By dividing this total sample into $d$ groups and calculating the mean of each group we can estimate the sampling error of the estimate.

For example in the multistage problem examined in the previous section we selected a sample of 300 units from a population of 100,000 units. If we wished to use replicated sampling to calculate the sampling error of the estimate we would proceed as follows.

1. Divide the total sample of 300 into 20 groups of 15 units each.
2. Allocate the sample of 300 units between the twenty groups as follows. The lot drawn by the first random number from the first branch inventory is allocated to group 1, the second to group 2, the third to group 3, etc., so that the units drawn from each branch are scattered among the 20 groups.
3. Calculate the mean of each group.
4. Calculate the mean, $m$, and standard deviation, $s$, of these group means—see Table 10.4.
5. Multiply $m$ by $N$, the number of units in the population, to arrive at a best estimate of the value of the population.
6. Multiply $s$ by the sample reliability factor from Table 10.1 to arrive at an estimate of unit sampling error. Multiply this last by $N$ to arrive at an estimate of the sampling error of the population.

The data are set out in Table 10.4. Since the unit mean and sampling error are £150 and £7·25 respectively the estimated value and sampling error of the estimate are

$$100{,}000 \times £150 = £15{,}000{,}000$$
$$100{,}000 \times £7{\cdot}25 = £725{,}000$$

We can state with a 90% level of confidence, that the actual value of the stores inventory lies between £14,275,000 and £15,725,000. We repeat that this is a very rough estimate and if more precise methods are available they should be used.

## QUESTION SERIES 10

1. What advantages might be derived from cluster sampling?
2. Traditional auditing uses what kind of sampling?
3. Why is a cluster sample likely to be less efficient than an unrestricted random sample of equal size?
4. Two thousand invoices are batched in groups of tens. You require a random sample of 300 units. You decide to use cluster sampling. How would you draw your sample? (*)
5. How do we calculate the mean and sampling error of a cluster sample?
6. Can a cluster sample be *more* efficient than an unrestricted random sample of equal size? (*)

7.

| | |
|---|---|
| Population size | 100,000 |
| Estimated population mean | £5·1 |
| No. of clusters | 50 of 30 units each |
| Required level of confidence | 95% |
| Standard deviation | £1·7 |

Calculate the precision limit on the estimate. (*)

8. The precision limit found in the previous question is too wide. A limit of £25,000 is required. Find a cluster sampling plan that will provide this precision limit. (*)

9.

| | |
|---|---|
| Population | 40,000 |
| No. of clusters | 20 × 10 units |
| Confidence level required | 99% |
| Cluster average | 30% |
| Required precision limit | ±4% |
| Standard deviation | 10% |

Calculate whether 20 clusters under these conditions gives the required precision limit. If not how many clusters are required? (*)

10. Suggest a multistage sampling plan for polling opinion before a General Election. (*)
11. If a three stage sample is drawn at how many points is sampling error introduced? (*)
12. Suggest an auditing situation from your own experience where multistage sampling might prove useful. (*)
13. Three steps are employed in multistage sampling. What are these three steps?
14. If $n$ units are drawn in a multistage sample will the inference be more or less accurate than the inference drawn from a pure random sample of equal size? (*)
15. If a complex sampling plan makes it difficult to calculate sampling error what alternative strategy is available?
16. Replicated sampling makes an important assumption about the population. What is this assumption? (*)
17. Set out the six steps required to calculate sampling error using replicated sampling.
18. What is the major difference between multistage sampling and stratified sampling? (*)
19. XY are a retail organization that operate sixty-four stores scattered over a wide area in Southern England. The group's internal auditor has reason to believe that substantial errors in pricing invoices have been made over the previous year by the stores. The 300,000 invoices represent a sales value of £30,000,000. He selects 10 random

stores and checks 50 invoices from each store. The results are as follows:

| Store | Mean value of error £ | Total no. of invoices | Estimated SD from samples |
|---|---|---|---|
| 1 | +4·7 | 5326 | 1·6 |
| 2 | +10·1 | 2170 | 5·2 |
| 3 | +6·3 | 10,830 | 2·3 |
| 4 | +0·1 | 6173 | 0·05 |
| 5 | +4·6 | 8774 | 1·7 |
| 6 | +5·7 | 2110 | 1·6 |
| 7 | +6·0 | 7347 | 2·3 |
| 8 | +2·3 | 1254 | 2·1 |
| 9 | +6·8 | 6133 | 1·9 |
| 10 | +9·4 | 876 | 6·5 |
| | +5·2 | 50,993 | 2·53 |

Estimate the total value of error and calculate a 95% confidence interval on this estimate. (a) Using conventional methods. (b) Using multistage sampling. (*)

20. A sample of 500 drawn from a complicated sampling plan is divided for replicated sampling as follows:

| Group | Mean value of unit in each group |
|---|---|
| 1 | 15·3 |
| 2 | 18·4 |
| 3 | 17·1 |
| 4 | 14·8 |
| 5 | 18·1 |
| 6 | 17·2 |
| 7 | 21·9 |
| 8 | 18·5 |
| 9 | 16·3 |
| 10 | 15·2 |
| | 17.3 |

Calculate a confidence interval for a 95% level of confidence in the estimate of the population value. The population size is 10,000. (*)

## SOME ANSWERS TO QUESTION SERIES 10

4. (a) Select thirty random numbers between 1 and 200.
   (b) Select the thirty batches corresponding to one of these numbers, i.e. the block of ten.
6. Yes, if the variability, say value, is greater within the cluster than between clusters. This is unlikely.
7. £510,000 ± £48,600.
8. Use Figure 21.5.4 on p. 252. 160 clusters approx.
9. No. You need about 45 clusters.
10. A judgement sample of representative constituencies. Then a random sample from the voters' roll.
11. At each of three stages.
12. Testing stock control procedures of retail stores spread over wide area.
14. Less accurate on the average.
16. The population is relatively homogeneous.
18. In stratified sampling the population is split up to improve the accuracy of the inference. In multistage sampling the population is split up to cut the cost of sampling.
19. (a) If multistage sampling is not used

| | |
|---|---|
| Population | 300,000 |
| SD | 2·53 |
| Confidence level | 95% |
| Sample size | 10 × 50 = 500 units |
| Precision limit | ? |

Looking up graph on p. 256 the precision limit is close to ± £70,000. However, this is not correct because the auditor has not sampled from a large section of the population.

(b) Using multistage sampling and so introducing sampling error at two stages.

1. SD of sample averages = 2·85
2. Average of square of sampling error from secondary sampling units = 7·351
3. No. of primary units = 64
4. No. of primary units samples = 10

Thus primary sampling error of estimate is:

$$2{\cdot}85 \times 0{\cdot}7533 = 2{\cdot}146 \qquad (2{\cdot}146)^2 = 4{\cdot}61 \text{ (See Table 10.1)}$$

Secondary sampling error of estimate squared is:

$$\frac{10}{64} \times 7{\cdot}351 = 1{\cdot}148$$

Total sampling error is therefore:

$$(4{\cdot}61 + 1{\cdot}148)^{1/2} = 2{\cdot}4.$$

The confidence limits are therefore:

$$£2{\cdot}4 \times 300{,}000 = £720{,}000$$

The estimate of total error value is:

$$300{,}000 \times £5{\cdot}2 = £1{,}560{,}000.$$

The 95% confidence interval is therefore.

$$£1{,}560{,}000 \pm £720{,}000$$

i.e.

$$£840{,}000 - £2{,}280{,}000$$

The auditor is now 97·5% sure that the total pricing error is not less than £840,000 on £30,000,000.

20. $(4{\cdot}16)^{1/2} = 2{\cdot}04$

Sampling error = 2·04 × 0·753 = 1·536
Estimate of value = 10,000 × 17·3 = 173,000
Confidence limit = 10,000 × 1·536 = 15,360

Conclusion: value is £173,000 ± £15,360 at 95% level of confidence.

# 11

# Sundry Topics

## INTRODUCTION

In this chapter I will elaborate on several aspects of audit sampling, which, although of some importance, do not justify a chapter to themselves. Several of the topics discussed have been touched on in earlier chapters, others are introduced for the first time.

## ONE TAILED TESTS, CONFIDENCE LEVEL AND SAMPLE SIZE

In Chapter 3 we defined confidence level as the probability of the population value falling within $\pm$ £$n$ of the estimated value. The confidence interval straddled the estimated value, in fact the estimated value lay at the mid-point of the confidence interval.

Figure 11.1(a) illustrates a situation where the area under the curve represents all of the sample estimates of the population value. 90% of the estimates lie within the precision limit $\pm$ £2000 of the estimated value of £$x$. In other words we can be 90% confident that the true population value lies $\pm$ £2000 either side of £$x$.

We notice that 10% of the values lie *outside* the precision limit of $\pm$ £2000. We also notice that *since the distribution is symmetrical*, 5% lie below £$(x - 2000)$ and 5% lie above £$(x + 2000)$.

Now suppose that it is of no concern to the auditor if the value of the population exceeds £$(x + 2000)$. This could be the case if £$x$ were the value of inventory and the auditor is only interested if there is a *deficiency* of inventory value. The auditor is only interested in the *lower* bound of the confidence interval.

In this case he can state that he is 95% confident that the population value exceeds £$(x - 2000)$.

By abolishing the upper bound of the confidence interval he has increased his level of confidence from 90% to 95% *without increasing sample size.*

This is a useful economy. But suppose the auditor only requires a 90% level of confidence. How does he discover the smaller sample size which will give him this level?

Clearly he must find that sample size which will give him an 80% level of confidence with a two tailed test. So long as the distribution is symmetrical this will give him a 90% level of confidence for a one tailed test.

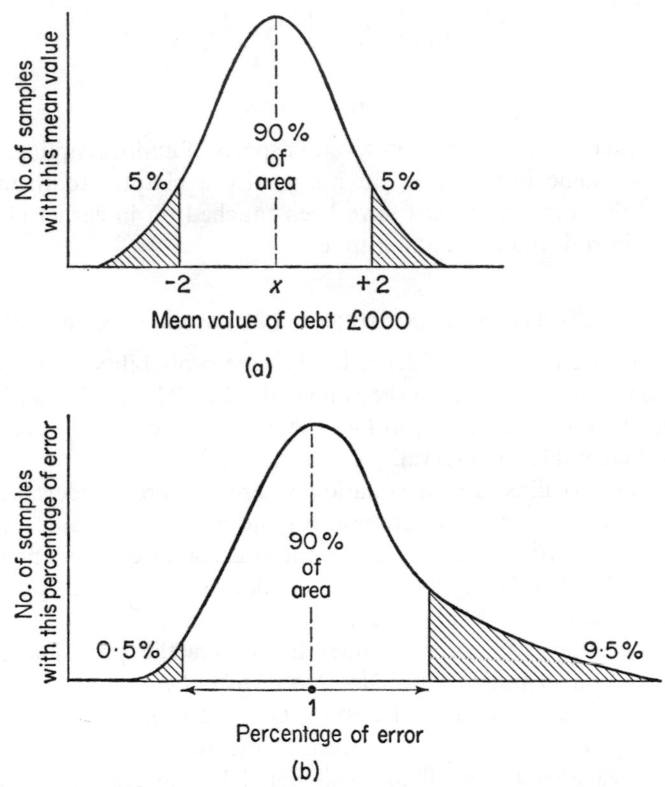


Fig. 11.1. Two populations of sample means. (a) From population of debts Normally distributed. (b) Population of error proportions of a series of samples drawn from a population containing 1% of error.

Let us take an example.

| | |
|---|---|
| No. of units | 10,000 |
| Estimated value | £50,000 |
| Precision limit | ± £1000 (£0·10) |
| Standard deviation | £2 |
| Level of confidence | 90% |
| Sample size for two tailed test | 980 units |

If the auditor wishes to transform this into a one tailed test he must find the sample size for an 80% level of confidence.

Tables for 80% level of confidence are not readily available so the auditor must use the formula outlined in Chapter 8, p. 80.

From the table of the Normal distribution on p. 238 we find that 80% of the area under the curve lies 1·28 standard deviations from the mean value ($Z = 1{\cdot}28$ for 0·3997 which is close to 0·4000).

Therefore:

$$1{\cdot}28s = 0{\cdot}10.$$
$$s = 0{\cdot}078$$

$$\frac{(S)^2}{(s)^2} = \text{required sample size}$$

$$\frac{(2)^2}{(0{\cdot}078)^2} = \frac{4}{0{\cdot}0061} = 656 \text{ units}$$

A two tailed test at an 80% level of confidence requires 656 units. If we abolish the upper bound we have an 80% + (100 − 80)/2 = 90% level of confidence that the population value exceeds £49,000.

This latter less rigorous precision may satisfy the auditor, and so he has achieved his objective despite a reduction of (980 − 656) = 324 in sample size.

## A CAUTION ON ONE TAILED TESTS

In the previous example the abolition of one side of the confidence interval increased the level of confidence by half the difference between the two tailed level and 100%.

That is:

| Two tailed | One tailed |
|---|---|
| 80% | equals 80 + (100 − 80)/2 = 90% |
| 90% | 90 + (100 − 90)/2 = 95% |
| 95% | 95 + (100 − 95)/2 = 97·5% |

But this only applies if the sampling distribution is symmetrical around the population mean. This, fortunately, is usually the case. If, however the sampling distribution is not symmetrical, because the population from which it is drawn is very highly skewed, the above reasoning does not apply.

Figure 11.1(b) illustrates a sampling distribution that is not symmetrical around the mean of $x\%$, the estimate of the error proportion in the population. In this case if we go out, say, 1·64 standard deviations from the mean value we find that, because the distribution is highly skewed, only 0·5% of the area lies below the lower bound and 9·5% lies above the upper bound. By abolishing the lower bound we increase our level of confidence by only 0·5% to 90·5% not by 5% to 95% as with the symmetrical distribution.

The graphs set out in Figures 21.1.1 to 21.1.3 allow for this lack of symmetry when we are working with small samples from highly skewed populations.

Figures 21.2.1 to 21.2.3 show both the upper and lower confidence intervals on the estimate. Note the lack of symmetry on low proportions.

This shows the asymmetry between the lengths of the two precision limits above and below the sample mean, i.e. (upper confidence limit − sample mean) − (sample mean − lower confidence limit).

We notice that the distribution is highly asymmetrical for very small proportions (or very large proportions!) and that the asymmetry diminishes as the error proportion approaches 50%. At 50% the distribution is symmetrical.

This suggests that when we are attempting to estimate a population proportion from a sample proportion, and the sample proportion is less than 10% we must be cautious in increasing our level of confidence by abolishing the lower bound of the confidence interval. This assumes that we are using a precision limit of equal length either side of the sample mean by using the standard error of a proportion formula.[1]

In these circumstances it is better to use specially calculated tables such as Arkin (1963), Table F to find the one tailed confidence interval.

However we must be careful not to overestimate the importance of skewness. Even when an accounting population is very highly skewed the population of sample means drawn from this population is relatively Normal so long as the sample is large, say over 500.

The auditor should beware where the expected proportion is 5% or less and the sample size is less than 500.

## SOME ADDITIONAL COMMENTS ON THE AVERAGE RANGE METHOD OF CALCULATING STANDARD DEVIATION

In Chapter 5 we described a quick method of estimating the standard deviation of an accounting population. The method used was to draw seven groups of seven units from the population, calculate the average

[1] See p. 88.

range of these seven groups, and divide this average range by a given constant, in this case 2·704 to arrive at an *estimate* of the standard deviation.

It is most important for the auditor to realize that the initial estimate from the preliminary sample of 49 is only a rough guide which must be checked later.

For example five groups of students were set to estimate the standard deviation from the following population by the author.

| | |
|---|---|
| No. of units | 1000 |
| Total value | £78,000 |
| Level of confidence required | 99% |

Using a random sample of 49 units divided into seven groups of seven the estimates of the five groups were as follows: £12·2, 15·4, 12·7, 16·4, 11·1, a difference of 32% between the highest and lowest estimate!

What effect would this have on the population size? The effect is shown below:

| (a) Unit precision limit | (b) Standard deviation | (c) Ratio (a)/(b) | (d) Sample size at 99% CL |
|---|---|---|---|
| £ | £ | | |
| 3 | 12·2 | 0·25 | 96 |
| 3 | 15·4 | 0·20 | 143 |
| 3 | 12·7 | 0·24 | 105 |
| 3 | 16·4 | 0·18 | 171 |
| 3 | 11·1 | 0·27 | 81 |

The largest sample size is 2·1 times the smallest. This fact, of course, affects the level of confidence in the inference. If the true standard deviation were 16·4 then the group who came up with 11·1 believed they had a level of confidence of 99% when, in fact, if we examine the tables, we find they had a level of confidence of only 90%. It is therefore important for the auditor to ensure that he does not *underestimate* the standard deviation. This means that he must recheck his estimate of the standard deviation from the full sample.

Grubbs and Weaver (1947) describe how a sampler can place precision limits on his estimate of the standard deviation. However, a much simpler, and quicker, method is to check the initial estimate from the full sample. Suppose the initial sample of 49 random units generates a standard deviation of 12·7. This suggests a sample size of 105. The auditor now has 15 groups of 7 and, *keeping them in the same order in which they are drawn*, he can calculate the average range from these 15 ranges and divide by 2·704 to calculate a more precise standard deviation. Alternatively he can regroup the initial sample of 49 drawn, if the sample size is 49. If the first grouping was 7 × 7 and divide by 2·704, the second grouping could be 8 × 6 and divide by 2·534. The auditor will then use the *larger* of the two standard deviations derived in this way to calculate the required sample size. See Table 21.6 for divisors.

Since the conservative principle applies to auditing, the auditor will not worry unduly if he overestimates the standard deviation since he is then underestimating his true level of confidence. He ought to be concerned if he underestimates the standard deviation since he is now overestimating his level of confidence in the inference.

By rechecking the standard deviation from the full sample the auditor reduces the latter probability.

The more groups of seven the auditor draws, the closer to the true standard deviation his estimate is likely to be. There is little point, however, in using more than around fifteen groups of seven. In other words, if the sample size is say 490, this could provide 70 groups of seven, but the auditor will only use the first 105 random units to recheck his estimate of the standard deviation.

In the above example we have used the magic number 2·704 for dividing the average range. This number depends upon the size of the group. The divisor for various group sizes is as follows:

| Units in group | Divisor |
|---|---|
| 6 | 2·5344 |
| 7 | 2·7043 |
| 8 | 2·8472 |
| 9 | 2·9700 |
| 10 | 3·0775 |

Figure 11.2 shows how this divisor is calculated. If, for example, a large number of groups of seven are drawn, the lowest and highest value in each group will tend to cluster around a mean value as shown in Figure 11.2. These mean values have been calculated by Grubbs and Weaver (1947) as standing 2·704 standard deviations apart. If we, then, estimate these mean values and divide by 2·704 we arrive at an estimate of the standard deviation! Notice that the larger the group the wider the gap between the mean values.

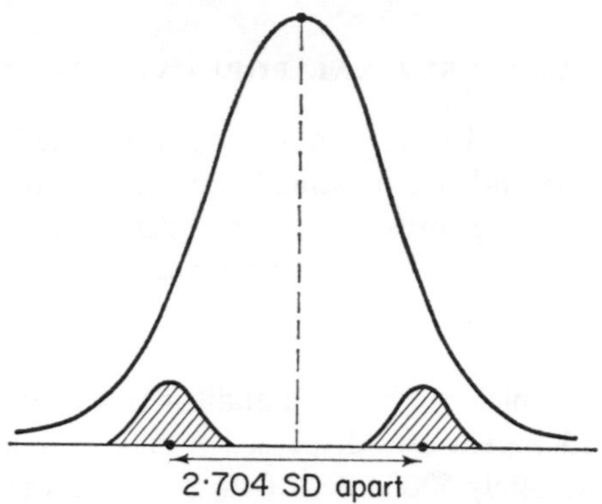


Fig. 11.2. The figure illustrates the principle behind the average range method of estimating standard deviation. Note that the means of the upper and lower bounds of the two sampling distributions are exactly 2·704 standard deviations apart.

To place confidence limits on the accuracy of the estimated standard deviation we need to know the standard error of the estimate. This latter figure depends on the size of the sample. The approximate standard error for various sample sizes can be found by using the coefficients:

| Sample size | Standard error coefficient[1] |
|---|---|
| 49 | 0·116 |
| 98 | 0·082 |
| 140 | 0·065 |

Thus if the average range comes out at 638 for 7 groups of 7 the most likely standard deviation is 638/2·7043 = 236. The 90% confident limits on this estimate are 236 ± 1·64 (0·116 × 236) = 236 ± 45, or 191 to 281.

The lower limit does not concern the auditor so the auditor can be 95% confident that the SD is below 281. However, this confidence interval is surely too wide? By increasing his sample size to 20 × 7 the auditor can reduce the precision limit to ±25 at a 90% level of confidence.

He can now state that he is 95% confident that the standard deviation does not exceed 261.

The above calculations are very approximate but they provide a rough method of ensuring that the estimate of standard deviation is not too low.

[1] Also see Table 3.4.

## ON TREATING SEVERAL POPULATIONS AS ONE

In Chapter 7 we explained how an auditor can reduce his sample size or increase his level of confidence by stratifying a population. A population which is split into several groups, say several branches, is already stratified, but this natural stratification may serve no useful purpose and may in fact prove very costly if the auditor decides to treat each section as a separate population.

Let us take an example. Suppose an auditor is testing the total value of inventory held in 15 warehouses dispersed over a wide area. Each warehouse holds approximately 5000 lines of stock so that the total population of lines is 15 × 5000 = 75,000. If the auditor treats each warehouse as a separate population and if the ratio of precision limit (acceptable sampling error) to standard deviation is 0·10, he will require to draw a sample of 258 lines from *each* warehouse to achieve a 90% level of confidence *in valuing each section.*

If however he treats the total population of 75,000 lines as a single population, then a sample of 270 is sufficient to provide the required level of confidence.

In the first case the sample size was 15 × 257 = 3855, in the second case it was only 270! A sample of 3855 from the 75,000 population would have provided a level of confidence of 99·99%!

We conclude that unless an auditor has reason to believe that a segregated population is *not* homogeneous with respect to error, etc., he should treat the several separate populations as a single population and this will often provide a massive reduction in sample size. The MUS system (Chapter 17) is based on this principle.

## ON CHECKING SEVERAL POPULATIONS FROM THE SAME SAMPLE

When an auditor is checking a sample from a population of documents it often happens that he checks several items on each document sampled.

For example, an auditor checking a sample of 200 invoices from a population of 10,000 invoices might check the following items on each invoice—date, authorization, price, extension, summation. Now each of these items is an *attribute* of the invoice and each group of attributes forms a separate population. Therefore the auditor is, in fact, testing five separate populations not one.

If the determinants of sample size are the same for each population and if the various populations are independent of one another no harm is done by using the same sample units to test each population and the benefit in saved sampling time is considerable. The reason why the same random sample can be used to test several populations is explained in a later section of this chapter.

This sanguine conclusion, however, only applies if the determinants of sample size, such as confidence level, standard deviation, population size, etc., are the same for each population. This need not be so. For example the auditor might consider that an incorrect authorization is more important than an incorrect price. Therefore he will require a higher level of confidence in testing authorizations than in testing price. Therefore the sample size is likely to be greater for authorizations than for price.

When this situation occurs the auditor will draw a sample size equal to the sample size required for the largest sample and use a fraction of this for the other samples.

Note that if the auditor draws 500 sampling units for the largest sample and a smaller sample requires, say, 100 units he can use 100 of the 500 units already selected. But, and it is an important but, he must use 100 units in the order in which they are drawn from the random number tables, or alternatively take a systematic sample of every fifth unit from the 500 units already selected. If he selected the first 100 of the 500 units once they had been sorted into sequence he would fail to sample from the last 80% of the invoice population!

A more serious problem arises if the several populations being tested from the same sample of invoices are not independent of one another. This problem can be best illustrated with an analogy.

In the latter half of the nineteenth century a French psychologist named M. Alphonse Bertillon developed a system called 'Bertillonage' which purported to be able to identify persons by making a list of their bodily measurements. He persuaded the French police to adopt the system for identifying criminals. M. Bertillon claimed that the probability of any two persons having the same set of measurements was exceedingly low.

However in England Francis Galton, the founder of Anthropometry, was highly sceptical of the 'extraordinarily large statistical claims' made by M. Bertillon. He considered that the unfortunate M. Bertillon had made a serious statistical error in ignoring the interdependence of the several variables.

'The incorrectness lay in treating the measures of different dimensions of the same person as if they were independent variables, which they are not. For example, a tall man is much more likely to have a long arm, foot, or finger than a short one. The chances against mistake had been overrated enormously owing to this error; still, the system was most ingenious and very interesting.'[1]

When an auditor checks several populations from the same sample he must beware against committing a similar error. If each population is prepared by a different individual it seems unlikely that an error in one part of the document is likely to induce an error in another part, but if several of the populations, say extension and summation, are carried out by the same individual an error in one part *might* induce an error in the other.

In the latter case the populations are not independent and, strictly speaking, a *different* sample should be drawn for each of these *interdependent* populations.

I do not think interdependence is very likely between accounting populations but the auditor can guard against it by adopting the following rule. If two or more errors are discovered on the same document check to see whether these items are prepared by the same individual or system. If so, draw an independent sample to double check one of the two potentially interdependent populations.

## A RANDOM SAMPLE IS A RANDOM SAMPLE IS A RANDOM SAMPLE

In the previous section we noted that the *same* random sample can be used to test several attributes of a population. This fact is of great significance to the auditor since it much reduces the cost of sampling. Since a random sample is an unbiased sample the same sample can be used again and again to test various aspects of the audit.

An auditor, for example, may draw a preliminary random sample to estimate the standard deviation of the population; this same sample can be used as the first $n$ units to estimate the error rate in the population, and the units in error can later be used to estimate the total value of error, and so on.

Long after an audit has been carried out a random sample drawn and noted in that audit can be used to make inferences about the population from which it is drawn. For example, if a fraud is uncovered after an audit is completed, the random sample can be used to calculate the probability of the auditor detecting *one* fraudulent unit from the sample he drew and also the sample size he would have needed to draw to be say, 90%, sure of picking up one fraudulent unit.

[1] Galton (1907).

This latter procedure might provide a useful defence strategy in a court of law.

## THE PROBLEM OF THE MISSING VOUCHER

The problem of dealing with an error of omission rather than an error of commission is an old one in auditing. How do you detect a missing voucher if you have no reason to suspect that it ought to be there in the first place?

The problem is covered in standard textbooks on auditing, and statistical sampling has nothing new to contribute towards a solution.

However if the auditor is vouching a set of payments and he finds that the voucher representing one of his random numbers is missing what does he do? There are two parts to the solution of this problem.

The missing voucher itself represents a matter requiring further investigation. In addition the omission of this voucher complicates the evaluation of amount or proportion.

I suggest that the best procedure for the auditor to adopt is as follows.

1. Substitute another random number for the one representing the missing voucher.
2. Complete the evaluation process using the substitute voucher.
3. Once (2) is completed tackle the problem of the missing voucher or vouchers as a separate problem requiring investigation.
4. Once (3) is solved compare the original missing voucher(s) value with the value of the substitute voucher(s). If the difference is significant the auditor may have to revise the inference he arrived at under step (2).

In my experience this last possibility is not likely to arise.

## THE FINITE POPULATION CORRECTION FACTOR

In Chapter 8 we noted that the standard error of a proportion is given by the formula:

$$s = \left(\frac{p(1-p)}{n}\right)^{1/2}$$

This formula provides an exact measure only if the proportion is 50% and the population concerned is of infinite size.

We have discussed the problem of estimating the reliability of estimates on skewed populations elsewhere. We will now elaborate on the problem of sampling from non-infinite populations.

No accounting population is of infinite size. Therefore the formula of the standard error given above will never be absolutely precise. If, however, the sample includes only a very small part of the population, say 1%, the population can be treated for all practical purposes as if it were infinite.

When the sample makes up a significant proportion of the population the confidence interval based on the standard error *can be reduced* by what is called the finite population correction factor (FPCF).

When the proportion of the population in the sample is small, say less than 2%, the FPCF coefficient is negligible being greater than 0·9900. Also, it is rather unusual for the sample from an accounting population to exceed 10% of the population.

The critical region then is between 2% and 10%.

Table 11.1 gives the finite population correction factor for samples including from 1% to 30% of the population.

Table 11.1. Table of coefficients of finite population correction factor. A close approximation of any other percentage between 1 and 30 can be found by linear interpolation, i.e. 28% = 0·8660 − (0·8660 − 0·8367) × 3/5 = 0·8484.

| If sample makes up $x$% of population | Multiply by this factor |
|---|---|
| 1 | 0·9950 |
| 5 | 0·9747 |
| 10 | 0·9487 |
| 15 | 0·9220 |
| 20 | 0·8944 |
| 25 | 0·8660 |
| 30 | 0·8367 |

Alternatively the reader can calculate the FPCF for himself using the formula

$$f = \left(1 - \frac{n}{N}\right)^{1/2}$$

where $f$ = FPCF, $n$ = number of units in the sample, $N$ = number of units in the population.

The formula will not, of course, operate for percentages in excess of 50%. We must take the reciprocal in such cases, i.e. for 80% calculate for 20%.

The significance of the FPCF can be gauged from the following examples:

| | |
|---|---|
| Population | 1000 |
| Proportion of sample *with condition* | 10% |

| Percentage of population in sample | Standard error ±% | 90% confidence interval (1·64) % |
|---|---|---|
| 0·10 | 30 | 0–59 |
| 1·00 | 9·5 | 0–26 |
| 5·00 | 4·2 | 3·1–16·9 |
| 10·00 | 3·0 | 5·1–14·9 |
| 20·00 | 2·1 | 6·6–13·4 |

However, the above confidence interval assumes an infinite population size. The population actually consists of 1000 units, therefore the FPCF and revised confidence intervals are:

| Percentage of population in sample | Standard error | FPCF | Revised 90% confidence interval |
|---|---|---|---|
| 0·10 | 30 | 0·9995 | 0–59 |
| 1·00 | 9·5 | 0·9950 | 0–26 |
| 5·00 | 4·2 | 0·9747 | 3·3–16·7 |
| 10·00 | 3·0 | 0·9487 | 5·3–14·7 |
| 20·00 | 2·1 | 0·8944 | 6·9–13·1 |

We conclude that, under normal circumstances, the finite population correction factor does not play a large part in determining accuracy of inference.

If the auditor ignores the FPCF he is being *conservative* in his estimates since the FPCF reduces the confidence interval (or sample size).

The above figures illustrate the fact that even when the sample contains 20% of the population the FPCF does not play a very large part in determining accuracy of inference.

At a 99% level of confidence with 10% of the population in the sample the comparative figures are:

| | Confidence interval % |
|---|---|
| 1. Ignoring FPCF | 2·26–17·74 |
| 2. Adjusting for FPCF | 2·66–17·34 |

Even at this level of confidence the effect is not large.

The reader should note that in all of the above examples we have ignored the skewness factor. The actual confidence interval at 99%, allowing for skewness, is 4% to 19·6% (see p. 246).

## ESTIMATING THE SIZE OF POPULATIONS

If the reader has followed the argument in the previous section he will understand that a precise calculation of population size is not needed for estimating proportions. The absolute size of population plays only a small part in determining accuracy of estimation of a proportion.

Suppose, for example, that an auditor draws a random sample of 200 from a population of what he supposes to be 2000 units. He finds 10% of the units to have the required condition so his confidence interval, at a 95% level of confidence, is 6·4% to 14·8%.

Later he discovers that the population size was actually 10,000, five times greater than he thought it to be. How does this affect his confidence interval? It does not affect it very much. His new confidence interval, *ceteris paribus*, is 6·3% to 15%.

If, however, the auditor is estimating a *value* rather than a proportion the previous argument does not hold. The auditor will estimate the mean value of a unit in the population and multiply this by the number of units in the population to arrive at total value. An incorrect estimate of population size will cause an error of exactly that proportion in the total value.

When using estimation sampling of variables the auditor must make a reasonably precise estimate of population size. In the majority of cases this presents no problem. The number of units in the population is known and can easily be checked. If the number of units is not known but the units are physically separate, say a tray of machine cards or a file of punched cards, the population size can be estimated very precisely by counting, say, the number of units per inch and multiplying this figure by the number of inches in the population.

Sometimes, however, it happens that the population being audited is dispersed over a number of documents and there are a variable number of units per document. For example the population might be printed on a set of statements with a variable number of invoices per statement.

In such cases we must use statistical methods to estimate the size of the population.

To take an example we draw a random sample of 50 statements and find that the number of invoices per statement varies from 1 to 30 with a mean of 3·4 and standard deviation of 0·6. If there are 3200 statements and we wish to estimate the population size to within ±2% of the actual figure we use our preliminary sample to calculate that we need a random sample of 290 statements to estimate population size to within ±2% of actual population size at a 95% level of confidence.

A larger sample would, of course, give us a more precise estimate of population size.

## ON CALCULATING THE LEVEL OF CONFIDENCE IN PREVIOUS YEARS WHEN STATISTICAL SAMPLING NOT EMPLOYED

The prime objective in statistical sampling is to relate the sample size tested by an auditor to the level of confidence he requires in the inference. It is of interest to examine this relationship in the years prior to the adoption of statistical sampling method. Was the sample size in previous years too high or too low? Or to put the same question another way, was the auditor asking for a level of confidence much higher or much lower than he really required?

The few studies I have carried out in this field have unearthed an astonishing discrepancy between one audit and another.

In one case, where the proportion of vouchers audited had remained constant for many years, the level of confidence had moved from about 97% to above 99·9%. The latter figure is surely too high?

In another case, where the auditor depended solely on depth vouching of only a few units the level of confidence was so low that I hesitate to disclose it.

The level of confidence, and precision limit, need not be the same for every audit but it should vary in the area of 80% to 99%, and it should be calculated rather than happen by chance.

The calculation of the level of confidence to be placed in a previous year's audit is complicated by the fact that most prestatistical samples are cluster samples and, as we note in Chapter 10, cluster samples *may* be less efficient than pure random samples of the same size. If, however, we ignore this complication the calculation of the confidence limits on previous years' audits is not difficult.

The parameters of accounting populations remain remarkably stable from year to year therefore we can use the standard deviation and precision limit of this year's population as an estimate of those of the previous year's population. The size of the previous year's population and the sample size are also, usually, easy to calculate. Therefore we have all of the information we need to calculate the previous year's level of confidence.

Let us take an example.

| Period | Population size | Sample size | To ensure that error rate less than | Level of confidence |
|---|---|---|---|---|
| 1 | 1000 | 200 | 1% | 90% |
| 2 | 1500 | 280 | 1% | 95% |
| 3 | 2000 | 400 | 1% | 99% |
| 4 | 10,000 | 1000 | 1% | 99·7% |

The auditor increases his sample size roughly in step with population size although when population size jumps from 2000 to 10,000 in one year good sense, or perhaps economic necessity, persuades him not to increase his sample size proportionately.

If the criteria for sample size is that the auditor wants to be 95% sure that if he finds no errors in the sample the error rate is below 1% then his sample size is not optimal in three of the four year periods. In period 1 it ought to be about 250, in period 3 about 290 and in period 4 about 300. The rapid increase in population size has induced the auditor to increase his sample size to too great an extent. This is the usual situation.

In the normal audit situation all factors are either known or estimated except sample size. In the present situation all factors are either known or estimated except confidence level. Therefore we use the tables in reverse. For example if the rate of precision limit to standard deviation is 0·10 on a population of size 2000 *and* if the sample size drawn were 500 we can calculate from Figure 21.5.3 on p. 251 that the level of confidence was 99%.

Similar calculations can provide the level of confidence in other cases. Note that a *precise* calculation is not required. If the tables tell us that the level of confidence is, say, far above 99% that is all we need to know.

## GRAPHS VERSUS TABLES

If an auditor wishes to calculate the sample size required to meet certain specified conditions he may:

1. Calculate the sample size from the given formula.
2. Consult a set of sampling tables.
3. Consult a sampling graph.

Certain large accounting firms in the United States have developed a 'programmed approach' to statistical sampling in auditing. Many of those methods use graphs rather than tables for determining sample size.

Perhaps because of this a myth has developed that the graphical determination of sample size is somehow superior to the tabular method. This belief is misplaced. The tabular and graphical methods provide the same answer in approximately the same time.

The above comment also applies to the use of *nomogramms* to determine sample size.

It is possible that the graphical method enjoys a slight advantage in that it provides the auditor with an overall composite picture of the determinants of sample size and this may assist him in choosing an optimal sampling plan.

For example Figure 21.5.3 on p. 251 provides a graph illustrating the sample size needed in estimation sampling of variables given (a) the unit precision limit, (b) the standard deviation, and (c) the required level of confidence.

Since the graph provides an overview of many possible combinations *at one and the same time* it helps the auditor to choose the most economical sampling plan to satisfy his requirements.

For example Figure 21.5.3 shows that to increase the level of confidence from 95% to 99% at a ratio below about 0·10 would be a very expensive decision.

Another possible advantage of graphs is that they are less open to error in preparation. Tables consist of thousands of figures, the calculation and transcription of these figures usually results in a few errors in even the most carefully checked tables. Since graphs are continuous, an error in any single figure is soon spotted and put right.

## WHAT DOES THE AUDITOR DO IF THE FEE IS INSUFFICIENT TO COVER THE REQUIRED SAMPLE SIZE?

An auditor examines a population, decides, after a good deal of thought, on the required sample size, and then calculates the cost, £*c*, of drawing and checking this sample size. He discovers that £*c* is above that portion of the audit fee which applies to this job. What does he do?

This question has wide implications which affect the relationship between the auditor, management, the shareholders and the government. I do not intend to open this particular hornets' nest at this time. I will limit my discussion to the immediate problem.

First the auditor should ask himself if the criteria he has used for sample size are not too severe. What are the consequences of his inference being wrong? If they are not serious could he not drop his level of confidence from, say, 95% to 90%?

If the answer to the last question is—'certainly not! I weighed up all these considerations before I decided on 95%'—then we must adopt a different approach. Can a more careful analysis of the population reduce our sample size?

Perhaps by stratifying the population the auditor can achieve his overall confidence level with a much smaller sample size? In my experience this is a common answer to the problem.

Another alternative is to use difference estimates rather than absolute estimates to test the value of a population.[1]

Several other methods of reducing sample size are suggested in this book. But supposing none of these are of any avail. What does the auditor do now?

There are only two options remaining. Either he reduces his sample size to fit the fee and accepts whatever level of confidence this provides, or he increases his fee to cover the cost of the increased sample size.

The answer to this question cuts to the very roots of modern audit practice. Is the auditor being paid by the shareholders to give the best opinion he can within the limits of a fee determined by tradition, or should an auditor charge a fee sufficient to satisfy a level of confidence in his judgement determined independently of the fee charged?

This important and difficult question must be resolved by the auditing profession before the question set at the head of this section can be answered. At the moment the profession *acts* as though it followed the former proposition yet it professes to follow the latter.

It is hardly surprising that the auditing profession finds itself the victim of so much criticism while this key question goes unresolved.

## USING STATISTICAL SAMPLING TO TEST SYSTEMS

Auditors test populations and clerical systems. A clerical system consists of a population of clerical operations performed between two points in time.

Statistical sampling can be used to supplement depth vouching in the audit of clerical systems.

Discovery sampling is particularly useful in checking that if *no* errors are found the auditor can be $x$% sure that the error rate is below $y$% with a $c$% level of confidence. The MUS system (Chapter 17) uses this approach.

[1] See Chapter 8. Also see Chapter 15 on the Bayesian approach.

Let us take an example.

A payroll procedure is being 'tested' using depth vouching. From first to last 76 clerical operations are performed on each payslip. The depth audit takes four payslips 'all the way through', from primary documents to final ledger entries.

The company employ 10,000 workers with variable conditions applying to different payrolls. The total number of clerical operations performed on payroll during the year is of the order of 40,000,000 operations!

The sheer size of this population seems daunting. Surely the auditor's level of confidence must be negligible if he only checks four payslips?

Fortunately, as we explained in Chapter 3 the accuracy of inference from a sample depends not upon the size of the population but upon the size of the sample.

The point is well illustrated in this problem. If a discovery sample of 325 units are checked from a population of 40,000,000 units and if no errors are found the auditor can conclude with a 95% level of confidence that the error rate in the population is less than 1%. This assumes that the errors are independent and randomly distributed.

By checking four payslips 'all the way through', the auditor has checked $4 \times 76 = 304$ operations. Thus if these operations are independent of one another, and no errors are found, the auditor has a rather better level of confidence in assuming a low error rate than one might suppose.

Whether all of the operations performed on a single payslip *are* independent of one another raises the problem of interdependence discussed in a previous section of this chapter. It is because I believe that the various entries on the same payslip are interdependent that I am sceptical about the value of depth vouching.

## USING THE COMPUTER TO ASSIST IN SAMPLING

Many companies today store their records on magnetic tape ready for processing on a digital computer. When the population of records to be audited, or a copy of them, is available on magnetic tape or disc the auditor can use this fact to speed up certain statistical sampling procedures.

The computer can be programmed to calculate, or at least estimate, the mean, standard deviation, and skewness of the population. 'Off the peg' utility programmes are almost certain to be available for this purpose. The computer can also draw the random sample of required size, and it can be programmed to calculate the sample mean and standard deviation. So far as generating random numbers is concerned it is better to feed a tested table of random numbers into the computer rather than rely on the computer generating its own set of random numbers. The latter would be untested and therefore suspect.

The computer can also assist in *stratifying* a population if the measures of standard deviation and skewness suggest that this procedure would be advantageous. The computer can also draw a judgement sample from the population, the attributes determining the sample being decided by the auditor. Notice that the computer can be programmed to carry out random and judgement sampling on the same run.

The analytical power of the computer is sometimes useful for carrying out certain tests on the population, for example, a sample of unit values can be converted to their logarithms to test whether they form a log–normal distribution.

Finally let us note the statistical sampling procedures which *cannot* be performed by the computer. The computer cannot decide the confidence level of precision limits suited to a given case; these depend on the auditor's judgement. Also the computer cannot perform the actual verification of the voucher, etc., chosen for audit, nor can the computer decide what to do next if the inference from the sample is unsatisfactory.

Several computer packages which have been devised to assist the auditor incorporate statistical sampling routines, i.e. AUDITAPE.

## SAMPLE SIZE AND THE INTERIM AUDIT

It is common for an auditor to visit his client several times in the course of a year and to carry out his audit in several stages. This raises the question of the sample size that should be tested at each stage of the audit. Let us take the simplest possible example.

An auditor carries out a yearly audit in two stages. The first stage is audited in July for the six months up to the end of June. The second stage is audited in December running into January for the six months to the end of December.

Suppose the auditor wishes to test check the invoices in July under the following conditions.

| | |
|---|---|
| Population to 30.6.?3 | 5000 |
| Estimated population six months to 31.12.?3 | 5000 |
| Confidence level | 95% |
| Precision limit | error rate less than 1% |

The sample size needed on 5000 units under these conditions is 300 and on 10,000 units is 305. If the auditor draws a sample of 300 for both halves of the year he achieves a 95% level of confidence in both halves and a 99·2% level of confidence in the whole, a higher confidence than he requires!

If, however, the auditor draws 300 units from the first half year and only 5 from the second he has breached the basic rule of random sampling, that each unit should have an equal chance of selection on each draw.

The auditor should therefore estimate the likely population size for the whole year at the interim audit, calculate the total sample size needed for this population and draw the requisite number of random numbers for the year at this time. If he uses a systematic sample he will draw a portion of the total sample from the first half year proportionate to the part of the population contained in the half year.

Returning to our example, the sample size for the full year is 305, the first half year is estimated to contain 50% of the population so the auditor draws a systematic sample of 153 from the first half year. This 153 gives the auditor a 95% level of confidence that the error rate in the first half year is *less than 2%*. But over both halves when added together he achieves his objective of less than 1% *if he finds no errors.*

Since population size has little effect on sample size it is not important if the auditor makes a wrong estimate of the total population size. Even a large misestimate will affect the sample size by only a very few units. This can be easily adjusted for by taking a few extra units in the first half year—just in case.

The key point in all this is that the auditor is assumed to be certifying the full year not the half year accounts. Therefore his level of confidence should be based on the population of the full year.

## QUESTION SERIES 11

1. The auditor has drawn a sample sufficient to give him a 95% level of confidence that the value of error lies between £20,000 and £40,000. He reflects that it is of no concern to himself if the error value is less than £20,000. State his level of confidence in the statement that 'the error value is less than £40,000'. Why would this assumption be wrong if the sample mean distribution was skewed 2% below £20,000, 3% above £40,000. What is his new level of confidence under these conditions? (*)
2. Give an auditing example where an auditor is not interested in (a) the lower bound of a confidence interval, (b) the upper bound.
3. Confidence level (two tailed) 50%, 80%, 90%, 99·7%. Calculate the one tailed confidence level. (*)

4. | | |
|---|---|
| No. of units | 5000 |
| Estimated value | £100,000 |
| Precision limit | ± £5000 (£1) |
| Standard deviation | £9 |
| Level of confidence | 90% (One tailed test) |

Calculate sample required. (*)

5. Why is the precision limit on a proportion only exactly symmetrical when the proportion is 50%?
6. What determines the relative accuracy of the standard deviation calculated by using the average range method? (*)
7. When estimating the standard deviation by the average range method using groups of seven why do we divide the average range by 2·704?
8. The SD is estimated from 7 groups of 7 to be 84. Calculate a 95% confidence interval on this estimate. (*)
9. The record keeping on debtors is decentralized by branch. Six branches have the following number of debtors

| | |
|---|---|
| A | 638 |
| B | 815 |
| C | 1121 |
| D | 634 |
| E | 192 |
| F | 1013 |
| | 4413 |

The auditor wishes to calculate the proportion of debts 6 months overdue for the 4413 accounts at a 90% confidence level within ±2% of the actual proportion.

What sample size does he need if he

(a) calculates each branch separately
(b) calculates only on the total of 4413?

The percentage is expected to be around 5%. (*)

10. What is the essential condition for treating several populations as one?
11. Why can the same random sample be used to test several populations?
12. If one job requires a sample of 300 and the next of 200 what important condition applies to selecting 200 of the initial 300 random units?
13. Why might a random sample drawn to test one population give a misleading result if used to test another population?
14. Suggest a rule for overcoming this last problem.

15. Suppose one of the random numbers drawn represents a missing voucher. What should an auditor do?
16. Calculate the finite population correction factor if the sample represents the following fractions of the population. 0·1%, 1%, 5%, 23%, 45%. (*)
17. Is the auditor taking a risk if he ignores the finite population correction factor? (*)
18. Why does an auditor not need a precise measure of population size if he is estimating a proportion?
19. How would you go about calculating the level of confidence in a previous year's audit when *statistical* sampling was not used?
20. In year 19?6 when statistical sampling was not used a sample of 300 units was drawn from a population of 5000 units. If the MUER now adopted is 1% what was the level of confidence in year 19?6. (*)
21. Suggest two advantages which graphs enjoy over tables for determining sample size.
22. If the cost of drawing the statistical sample is higher than the portion of the audit fee due to this job, suggest three strategies that the auditor might employ.
23. In a depth audit the auditor checks thirty invoices 'all the way through'. Twenty operations are performed on each invoice. There are 40,000 invoices in the population. What is his level of confidence that the clerical error rate on invoices is less than 1% if he finds no errors? (*)
24. State three sampling procedures that can, and three that cannot, be performed by a computer.

## SOME ANSWERS TO QUESTION SERIES 11

1. (a) 97·5% (b) 97%.
3. 75%, 90%, 95%, 99·85%. Always assuming the distribution of sample means is symmetrical around the mean value.
4. We must calculate the 80% level of confidence on a two tailed test

$$1{\cdot}28\ s = 1{\cdot}00$$
$$s = 0{\cdot}78$$

Thus

$$\text{sample size} = \frac{(S)^2}{(s)^2} = \frac{81}{0{\cdot}608} = 133$$

Therefore 133 units gives a 90% level of confidence on a one tailed test. Since we have assumed an infinite population size this is a conservative estimate.

6. The number of groups used to calculate the average range.
8. $84 \pm 19$
9. (a) 220, 229, 249, 220, 110, 243 = 1271
   (b) 300
16. 0·9995, 0·9950, 0·9746, 0·8775, 0·7416
17. No. He is being conservative.
20. The level of confidence was around 95% if we assume the example to be a random sample, and no errors were found.
23. Above 99%, but only if we make the improbable assumption that all of the populations audited on the invoices were independent of one another.

# 12

# The Detection of Fraud

## INTRODUCTION

An erroneous unit in an accounting population may be entered accidentally or on purpose. In the latter case we use the term fraud.

The major difference between fraud and accidental error is that the misfeant will attempt to outwit the auditor by placing the erroneous unit where he believes it is least likely to be detected. Accidental errors, on the other hand, are randomly scattered throughout the population.[1]

We noted above that one of the advantages of random sampling is that no-one, not even the auditor himself, is able to predict in advance the unit of the accounting population which he will audit.

This prevents a potential misfeant from discovering a pattern in the audit, i.e. never auditing the same month two years running, and using this fact to perpetrate a successful fraud. Smurthwaite (1965) has discovered such a pattern in audit work.

Perhaps we should mention at this point that the value of money taken by fraud is running at more than double the value of goods stolen in the UK in recent years.[2]

But can statistical sampling contribute more to the detection of fraud than randomizing the units selected for audit? The answer, as we shall see, depends upon the type of fraud perpetrated.

## LEVELS OF SIGNIFICANCE

An audit clerk might be introduced to an audit with the words 'Ignore everything under £10,000 and don't waste much time on items under £50,000.'

In this case amounts under £10,000 are not *significant* to the level of audit at which he is working.

In every audit there is some level of £$x$ below which amounts are not significant. We do not audit individual items below £$x$ *we audit the system by which these populations are controlled.*

[1] Although there is some evidence that accidental errors tend to cluster.

[2] See London Commissioner of Police (1973).

In most audits there are three levels of significance. The external audit level, the internal audit level and the level of the cashier. We might postulate £1000, £100 and £0, as the significant amounts at each of these levels. In real life there may be more than three levels and the amounts we have chosen as significant at each level are, of course, quite arbitrary.

For a fraud to be significant to the external or internal auditor the amount involved must equal or exceed the significant value for that level. For example, in the above case, the fraud must exceed £1000 *in total value* to be significant to the external auditor.

Table 12.1 shows a variety of ways in which a fraud of £1000 can be brought about. The misfeant could manipulate a 'big-bite' fraud, a single fraudulent unit of £1000. At the other extreme he might manipulate what has been called a 'salami' fraud. That is taking a relatively small amount a large number of times.

Digital computers are ideally suited to the 'salami' type fraud. Several years ago, in California a computer programmer rounded down all of the wages except one in a 100,000 man payroll that was supposed to be rounded to the nearest cent. The balance of 'round ups' he transferred to a colleague's payslip.

If we suppose that a multi-unit fraud is so interdependent that by detecting one fraudulent unit we will be led on to the next, what is the probability of picking up the fraud at various sample sizes?

## THE NEEDLE IN A HAYSTACK PROBLEM

When a fraud is confined to one or a few units of a *large* population no method of random sampling will give the auditor much of a chance of picking up the fraud.

Take, for example, the situation where an auditor is testing a population of 10,000 vouchers. The probability of picking up one defective unit, if one unit from the 10,000 is defective, is as follows:

| Size of random sample | Probability of finding *one* error in sample |
|---|---|
| 100 | 1% |
| 1000 | 10% |
| 5000 | 50% |
| 9000 | 90% |

Even when the number of errors increases to 10 the sample size has to be very large to give the auditor a reasonable probability of picking up *a single example* of the error, i.e.

| Size of random sample | Probability of finding *one* error in sample |
|---|---|
| 100 | 9·6% |
| 500 | 40·1% |
| 1000 | 65·1% |
| 2000 | 89·3% |
| 3000 | 97·2% |

If we accept the conventional confidence level of 90% the auditor would have to check a random sample of 2000 units to have a 90% chance of picking up *one* error. If 10 defective units occur in various population sizes the sample size needed to achieve a confidence level of 90% in finding one defective is as follows:

| Population size | Sample size needed |
|---|---|
| 2000 | 400 |
| 5000 | 1000 |
| 10,000 | 2000 |
| 1,000,000 | 20,000 |

The sample size is a constant 20% of the population size.

Statistical sampling then is not of much use for finding errors which constitute a very small proportion of the population.[1]

## THE 'SALAMI TYPE' FRAUD

As fraud moves towards the other end of the spectrum the probability of finding *at least one* fraudulent unit rapidly increases.

Table 12.1 sets out the sample size required to provide a 90% level of confidence of picking up at least one fraudulent unit for various percentages of fraudulent units. Above a fraudulent percentage of 0·4% the sample size begins to approach the range of sample size we use in detecting accidental error. The question the auditor must ask himself is this.

[1] Nor, incidentally, is any other method of sampling. If, however, the auditor has supplementary information about the likely areas of fraud he can *stratify* the population and so increase the probability of finding fraud in the suspect stratum.

'Is it worth while increasing my sample size from the normal range of 200–600 units in order to improve my chances of detecting fraud?'

I think the auditor can best answer this question by calculating the *expected value* of the fraud.

I asked several large professional accounting firms in what proportion of their audits (1) they had detected some kind of fraud or (2) fraud had been discovered after the audit was completed. In every case the proportion was well below 1% of audits.

Table 12.1. Sample size needed to provide approximately 90% level of confidence that at least ONE fraudulent unit from group of fraudulent units is detected. Population of 10,000.

| No. of fraudulent units | Value of fraudulent units £ | Percentage of population which is fraudulent % | Sample size required to pick up ONE fraudulent unit with 90% level of confidence |
|---|---|---|---|
| 1 | 1000 | 0·01 | 9000 |
| 5 | 200 | 0·05 | 3300 |
| 10 | 100 | 0·10 | 2000 |
| 40 | 25 | 0·40 | 600 |
| 50 | 20 | 0·50 | 500 |
| 100 | 10 | 1·00 | 220 |
| 200 | 5 | 2·00 | 120 |

Let us, then, assume that, unless the auditor has prior information suggesting fraud, the probability of fraudulent units being present in a population is no greater than 1%.

Thus the expected value, $v$, of the fraud is $£f \times p = £v$ when $f$ is the maximum value of fraud and $p$ the probability of fraud. If for example the auditor checks all units exceeding £10,000 the expected value of the single fraud is at maximum $£10{,}000 \times 0{\cdot}01 = £100$. If the cost of sampling one unit of an accounting population is £0·25, then it would not pay the auditor to sample in excess of 400 units since $400 \times 0{\cdot}25 = £100$ the expected value of the fraud.

It is likely that the larger the audit the more rigorous the internal audit control procedures and so the less likely fraud becomes. Thus as $f$ increases $p$ is likely to diminish.

If the reader accepts the logic of the previous argument we can conclude that the order of sample size used to locate accidental error, say 200 to 600 units, is approximately equal to the size of sample it will pay the auditor to draw to detect fraud.

Note that throughout the previous discussion we have assumed

1. That a single fraud has been perpetrated.
2. That the auditor checks all units of accounting populations above a certain value.
3. That where a fraud is made up of several units, the detection of one unit is sufficient to draw attention to the remaining units.

These conditions do not seem to us to be unreasonable.

## SUMMARY

If we assume that an auditor stratifies the population to be audited and carries out a 100% check on all units in the top value stratum equal to or above £$x$, then the problem of detecting fraud is limited to sampling from the units below £$x$.

The probability of detecting fraud obviously depends upon the number of fraudulent units. When these exceed about 0·5% of the population, conventional sample sizes for detecting accidental error will give the auditor a reasonable level of confidence of detecting *at least one* fraudulent unit.

When the number of fraudulent units are very small the auditor faces the 'needle in a haystack' problem and no amount of sampling ingenuity is likely to give him a reasonable level of confidence in detecting the fraud. We argue, however, that since the expected value of the fraud is not high it is improbable that it will pay the auditor to increase his sample size simply to improve his chances of detecting fraud.

## QUESTION SERIES 12

1. Why does random sampling assist in preventing fraud?
2. If a significant value in a given audit is £5000, how do we test for fraud above £5000 and below £5000?
3. If there are ten fraudulent units in a population of 2000, use Table 21.2 on p. 239 to calculate the sample size required to find one of these twelve units at a 90% level of confidence. (*)
4. The same problem as (3) except that:

| Population size | Percentage of fraudulent units |
|---|---|
| 1,000,000 | 0·5% |
| 50,000 | 1·0% |
| 10,000 | 2·00% |

(*)

5. If the maximum value of a fraud is £50,000 and the likelihood that it will occur is 0·1%, what is the expected value of the fraud? If it costs £0·25 to sample and check one unit, what is the maximum size sample it is worth while drawing to try to detect the fraud? (*)
6. 'If statistical sampling cannot find a "needle in the haystack" type of fraud it is useless as an audit tool.' Do you agree? (*)

## SOME ANSWERS TO QUESTION SERIES 12

3. 410.
4. 475, 240, 114.
5. £50, 200.
6. No, the main purpose of auditing is not to find fraud. Other sampling methods will provide no better chance if no more supplementary information is available.

# 13

# Case Study on Statistical Sampling in Auditing—Verifying the Provision for Doubtful Debts

Axel Finance has operated a hire purchase credit scheme for many years. In 19?6 the company set aside £120,000 for doubtful debts on a total debt of £2,412,000.

Axel Finance has up to date been audited by S. a rather small professional firm, but in 19?5 a major supplier of finance to Axel had insisted on L. an international firm of accountants sharing the audit. An audit senior from L., Mr. S.A., carries out a fact finding survey on Axel. He decides to check on the doubtful debts position. One of the key statistics in this type of business.

From 19?1 to 19?4 the doubtful debts provision had been calculated by checking through the year end accounts and listing all accounts where two payments have not been made out of the last four payments. Any account with a worse record than this is listed as 'bad' and transferred to a special ledger.

In 19?5 the chief accountant had declined to carry out this laborious procedure for calculating doubtful debts. He argued as follows:

'Over the last ten years the doubtful debts position has varied between 3% and 6% of total debts. Subsequent checks by my staff have shown that in no year has more than 6% of debts, in value, ever gone bad. Therefore I can save my department a lot of trouble by taking the bad debts provision to be 5% of the year end value of total debt'.

In 19?5 and 19?6 Axel have followed this policy in estimating the doubtful debts provision. The audit manager of L., however, is not happy about this procedure.

The key statistics in round figures on Axel's debt position over the last six years is as follows:

| | No. of a/c | Total value (£1000) | Doubtful debt provision (£1000) |
|---|---|---|---|
| 19?1 | 32,000 | 920 | 40 (38)[1] |
| 19?2 | 36,000 | 1000 | 52 (53) |
| 19?3 | 38,000 | 1100 | 54 (57) |
| 19?4 | 41,000 | 1300 | 62 (60) |
| 19?5 | 52,000 | 1900 | (95)(101) |
| 19?6 | 60,000 | 2400 | (120) |

In recent years Axel has expanded from financing 105 dealers in 19?3 to financing 153 dealers in 19?6. Most of this expansion has been in new town areas, and the audit senior of L., Mr. S.A., knows from his experience of other audits that these new towns present greater credit risks than the areas in which Axel had previously operated. He finds that the doubtful debts provision is split as follows:

Old town debts £80,000
New town debts £40,000

He is also worried about the exceedingly rapid growth of business in the last two years when the economy as a whole has been relatively stagnant. Have dealers been taking greater risks to get business?

Mr. S.A. puts these questions to Mr. M.D., the managing director of Axel, who was appointed in 19?4. Mr. M.D. claims that he 'monitors dealers most carefully' and that Mr. S.A. 'has nothing to worry about'. 'The quality of our customers today is better than it has ever been', states Mr. M.D.—and invites Mr. S.A. to lunch at the Hilton Tower.

Mr. S.A.'s next step is to inspect the accounting system for controlling credit. He has been warned by a colleague in S. that 'The bookkeeping system must have been put in by Paccioli'—and he finds no grounds for disagreeing with this verdict.

Axel have centralized their hire purchase records in an office in the centre of London. The individual accounts are kept on a set of ledger cards which are regularly updated by bookkeeping machine. However, many corrections are made by hand, in ink, and even occasionally in pencil. Mr. S.A. examines a few accounts and finds them not at all easy to follow. Dates are sometimes not sequential and corrections are often squeezed in between typed entries.

[1] The figures in brackets represent the actual bad debts written off during the following year.

Thoroughly alarmed Mr. S.A. arranges a meeting with the partner in L., Mr. P., who is supervising this audit.

P. decides to put a senior clerk into Axel who will go through the accounts and note the doubtful debts—using Axel's own criterion of a doubtful debt.

After this clerk has been working for five days Mr. S.A. drops into to see how he is getting on. The clerk has only been able to check 150 accounts out of the 60,000.

A hurried conference is held and it is decided to call in Mr. M.S. from the management services division who is an expert on statistical sampling.

Mr. M.S. asks to see the result of the sample of 150 accounts which have already been checked. These are the first 150 accounts so they may not be a true random sample, but they are better than nothing.

The results are:

| | |
|---|---|
| No. of accounts | 150 |
| No. doubtful | 12 |
| Percentage doubtful | 8% |
| Average value of all 150 | £40·3 |

Mr. M.S. calculates that the standard error on 8% for a sample of 150 is 2·21. Therefore 5% falls within the 95% confidence limits of 8% ±4·3%. The client's estimate of a 5% proportion falls within this confidence interval and is, therefore, not disproved.

Mr. M.S. asks if the 150 accounts can be divided into two groups. Group O being the accounts of customers in the older areas, Group N being the accounts of customers in the New Town areas. He is told that this can be easily effected using the account numbers which code the areas.

The resulting breakdown is as follows:

| | Old town | New town | Total |
|---|---|---|---|
| Population size | 40,000 | 20,000 | 60,000 |
| No. of accounts | 99 | 51 | 150 |
| No. doubtful | 4 | 8 | 12 |
| Percentage doubtful | 4% | 16% | 8% |
| Average value of all accounts | £35·1 | £50·4 | £40·3 |

5% is outside the 95% confidence interval of 16% ± 10%, so Mr. M.S. advises Mr. S.A. to concentrate on checking the value of the New Town doubtful debts. The 5% provision is almost certainly adequate for the

Old Town doubtful debts, since both the sample proportion and average value are below the provision provided.

Mr. M.S. decides that he wants to be 95% sure that he estimates the doubtful debts proportion for the New Town debts to within ±4% of the actual proportion. Therefore he needs to calculate the sample size $n$ in the formula.

$$e = \left\{\frac{p(1 - p)}{n}\right\}^{1/2}$$

when $e$ = the standard error of the proportion,[1] $p$ = expected percentage doubtful (i.e. 16%), $n$ = required sample size.

$$2{\cdot}04 = \left\{\frac{16(100 - 16)}{n}\right\}^{1/2}$$

$$n = 323$$

Mr. M.S. examines the method of allocating account numbers and concludes that he has no reason to suspect that the first 150 accounts are not a random selection. Therefore he puts two audit clerks onto the job of drawing a further 272 random New Town accounts. This takes another four days.

The cumulative results of the two draws are as follows:

| | |
|---|---|
| Population size | 20,000 |
| No. of accounts checked | 323 |
| No. of doubtful debts | 43 |
| Percentage doubtful | 13·3% |
| Average value of 323 debts | £50·6 |

Mr. M.S. concludes that he is 97·5% (i.e. 95 + (100 − 95)/2)% sure that the proportion of New Town debts which are doubtful is at least 9·3% (i.e. 13·3% − 4%).

Mr. M.S. decides that he wants a confidence limit of ±4% on both estimates of proportion.

In the case of the Old Town accounts since the estimated proportion, 4%, is less than 10% he cannot use the formula for standard error set out above. He must consult a suitable set of tables or graphs. He consults Figure 21.1.2 at the end of this book and finds that if he draws a sample of 140 and finds 4% doubtful he can be 95% sure that the proportion of doubtful debts in the population is 8% or less. He already has a random

[1] $e$ is calculated as follows: $1{\cdot}96 \times e = 4$, thus $e = 2{\cdot}04$.

sample of 99 Old Town accounts so he draws another 41 accounts. The percentage doubtful remains at 4%.

The average value of all accounts can be calculated by dividing the actual total debt owed to Old and New Town, which is known, by the actual number of accounts owed. The answer is

| | Total value (£) | No. | Average value (£) |
|---|---|---|---|
| Old | 1,404,000 | 40,000 | 35.1 |
| New | 1,008,000 | 20,000 | 50·4 |
| | 2,412,000 | 60,000 | |

But is the average value of doubtful debts the same as the average value of all debts?

The figures are as follows:

| | No. of doubtful debts | Average values (£) | Estimated standard deviation (£) |
|---|---|---|---|
| Old | 6 | 37·2 | 15·1 |
| New | 43 | 48·1 | 21·3 |

90% confidence limits on £48·1 with a sample of 43 is £5·25 (See Chapter 8, page 79 for formula).

This embraces the rather wide confidence interval of £42·85 to £53·35. Past records of the Old Town debts indicate that the average value of doubtful debts is not significantly different from that of all debts. Therefore, since £50·4 falls within the confidence limits and since it is above £48·1, Mr. M.S. decides to use the figure £50·4 as the average for New Town doubtful debts.

The number of Old Town doubtful debts are so small that it is not worthwhile calculating a confidence interval. He decides to use the method adopted in previous years and takes the average value of all Old Town debts, £35·1, as the average value of doubtful debts.

'We have collected lots of figures', says Mr. S.A., 'But what do they mean?'

Mr. M.S. hands Mr. S.A. a sheet of paper containing some calculations; this is reproduced in Table 13.1. Mr. M.S. says, 'If we assume that the average value of the doubtful debts are not significantly different from the overall debt average, an assumption not disproved by the available evidence, then we have placed 95% confidence limits on our estimate of the value of doubtful debts.' These are shown in Table 13.1.

'What effect does this have on the existing provisions?' asks Mr. S.A. 'The Old Town provision may be a little on the high side', replies Mr. M.S. 'but notice that our estimate based on the sample data gives a 5% probability that the Old Town debt exceeds £112,000! However there is a 74% chance that it is less than £80,000 so I think that £80,000 is a reasonable provision.

Table 13.1. Placing confidence limits on the value of Old and New Town debts.

*Estimate of Old Town debts*

We are 95% confident that the population percentage does not exceed 8%. Therefore

| | Most likely | Maximum |
|---|---|---|
| Population % | 4 | 8 |
| No. of debts | 1600 | 3200 |
| Average value | 35·1 | 35·1 |
| Total value (£) | 56,160 | 112,320 |

*Estimate of New Town debts*

We are 95% confident that the population percentage lies between 9·3% and 17·3%. Therefore

| | Minimum | Most likely | Maximum |
|---|---|---|---|
| Population % | 9·3 | 13·3 | 17·3 |
| No. of debts | 1860 | 2660 | 3460 |
| Average value | 50·4 | 50·4 | 50·4 |
| Total value (£) | 93744 | 134,064 | 174,384 |

'The New Town debt is clearly hopelessly inadequate. £40,000 represents a percentage of 800 doubtful debts, only 4% of the population! This figure lies 4·56 standard errors below the sample average of 13·3%. A probability so remote as to be virtually impossible. No, I am 97·5% sure that the doubtful debts provision on New Town accounts should be at least £94,000 and I would prefer £130,000.'

The following day a meeting is held between Mr. M.S., Mr. S.A., and Mr. P., the supervising partner in L.

'I have examined your estimates of the New Town doubtful debts', says Mr. P., 'and so far as I can understand your calculations your estimate seems reasonable. We must, however, be very cautious because I happen to know that the Board of Axel are negotiating the sale of their business and the buyer is waiting for the audited accounts. If we insist on raising the doubtful debt provision from £120,000 to £210,000, this will reduce profit by 20% and almost certainly influence the price the buyer is willing to pay. How sure are you of your estimate?'

Mr. M.S. replies 'we are accepting Axel's own estimate on the two-thirds of accounts which we have called Old Town debts. We disagree only on the provision on New Town debts.

'In the past the average value of doubtful debts has not differed significantly from all debts. The evidence we have about the doubtful debts on New Town accounts does not disprove this contention. Therefore we have assumed that the average value of New Town debts is £50·4. The only unknown then is the proportion of New Town debts which are doubtful. We accept Axel's own definition of a doubtful debt.

'By using statistical methods we can say that we are 95% confident that the proportion of doubtful debts lies between 9·3% and 17·3% with a most likely figure of 13·3%. This means that taking the average value as £50·4, we are 95% confident that the value of New Town debts lies between approximately £94,000 and £175,000. We are, in fact, 97·5% confident that the value is at least £94,000. This virtually disproves Axel's estimate of £40,000.'

Mr. P. says 'If I insist on an estimate of at least £180,000 overall I am well covered.' 'Most certainly,' replies Mr. M.S.

'If a 100% check of all New Town accounts is carried out there is only about one chance in thirty that the value of doubtful debts will turn out to be less than £100,000.'

'Thank you,' says Mr. P., 'you have both been most helpful. I will put this matter before the next partnership meeting. There are, of course, several other important points to consider, such as our obligation to the buyer and the possibility of the Old Town debt provision being excessive, but your minimum estimate of the New Town debt provision is the key figure in this affair.'

# 14

# Statistical Cost Allocation

## INTRODUCTION[1]

The major contribution of statistical sampling to accounting is in the area of audit and control. There are, however, other areas of accounting where the method is of value.

In this chapter we will examine the application of statistical sampling to one of these other areas. The particular area examined is that of cost allocation.

## THE COST ALLOCATION PROBLEM

Conventionally we separate allocated cost into two classes, *direct* and *indirect*. Direct cost can be allocated between projects without argument as to the basis of allocation. Indirect cost must be allocated on some arbitrary basis such as floor area or number of employees. The value of the latter procedure is open to question but it will not concern us in this chapter. We are solely concerned with the allocation of direct cost.

Why do we allocate direct cost? We suggest two motives. One concerns the decision function, the other the control function. We define *contribution* as the difference between the revenue derived from a project and its incremental cost. In many cases, direct cost provides a close approximation of incremental cost, and revenue less direct cost is taken as a measure of the contribution of the project to overall company profit. Thus direct cost is an input to the decision on whether or not to implement a project or process or continue to produce a given product.

The second reason for allocating direct cost is to prevent the misuse of resources.

The conventional method of preventing the misuse of a resource is to charge the direct cost of the resource against the user department. It might be argued that incremented cost and not direct cost should be charged but this is not common practice. Thus, direct cost becomes an input to the procedure for controlling resource utilization.

[1] An earlier version of this chapter appeared in *Accountancy*, **81**, 918, February 1970, 101–106.

We conclude that most accounting departments in business spend a significant proportion of their time on the activity of allocating direct costs between jobs, projects or processes. The resources utilized on this activity are likely to have a significant opportunity cost.

## THE TRADITIONAL ALLOCATION PROCEDURE

The conventional method of allocating direct cost is to open an account for each activity and charge each account in proportion to the amount of the total resource utilized. The resource might be either a product or a service. If it is a product, any unutilized balance will be charged to the inventory account at the end of the period. In the case of a service, the unutilized balance is a sunk cost and might be debited direct to manufacturing p and l account.

On the data processing side the allocation of direct cost is usually effected by the distributing agent filling up a slip or punched card and forwarding this to the costing department. One or more collating procedures may intervene between the initial recording and the final entry in the job account. Even if the final allocation is handled by a digital computer, the *initial* recording is usually performed by a human agent.

If a resource is used many times by a large number of users, if, for example, a maintenance department performs several thousand jobs a month or materials are allocated in many small batches, the data processing cost can be significant. Note that it is the number of transactions, not the *value* of the transactions that determines the data processing cost.

If the cost of allocating direct cost is significant, it may be worth while examining the comparative costs of the various methods of performing this operation. We suggest three methods:

1. Full direct cost allocation by manual methods,
2. Full direct cost allocation using EDP equipment,
3. Statistical cost allocation.

We will argue that under certain conditions statistical cost allocation is the best method to use.

## MANUAL VERSUS MACHINE METHODS

The choice between manual and machine methods depends upon the level of activity, but since labour is becoming steadily more expensive relative to other goods, the break-even point for manual methods is moving gradually to the left. Manual methods are only viable at low levels of activity.

All data processing methods using machines have a high set-up cost plus, in the case of unit record systems, a high input cost. These comparative disadvantages over manual systems must be compensated for by exceedingly low processing and output costs. However, as we noted above, if the input to a machine process is a joint cost with some other job which *must* be performed in any case, then the incremental cost of machine processing is very competitive and it is improbable that any other method of allocation will prove more efficient.

We conclude that where the input to a direct cost allocation procedure is a joint input with some other necessary job, machine processing is likely to prove the better option.

There are, however, a wide variety of direct cost allocation applications where the necessary condition of joint input does not apply. Under these conditions, we suggest that statistical cost allocation should be considered.

## STATISTICAL COST ALLOCATION

The basic idea in statistical cost allocation is to draw a sample from the population of items to be allocated. For reasons which we will explain later, the sample may be as small as 5% of the total population or even smaller if the population is very large. Clearly, this method will bring a considerable reduction in clerical costs over other manual methods, and, as we noted above, it may also enjoy advantages over machine methods.

Statistical cost allocation can be adapted to a wide range of circumstances, but the ideal conditions for its use are as follows:

1. A large population of costs are to be allocated. At least several hundred a month and preferably, several thousand.
2. None of the costs in the population should be *individually* of significant value. Or, alternatively, prior information should exist to enable the sampler to stratify the population to remove individual items of significant value. The decision on what amount is 'significant' depends on the judgement of the accountant, etc., under the given conditions.
3. The allocation procedure should not require absolute accuracy. For the type of decision taking and control purposes outlined above, *absolute* accuracy is seldom required.
4. The resources used in operating the existing allocation procedure should have a significant opportunity cost.
5. The documents employed in the allocation procedure should be suitably arranged for statistical sampling purposes. It is unlikely that they will *not* be, but if they are geographically dispersed in small

batches, some technical sampling problems might arise. Note that the documents need not be *numbered consecutively*, or numbered at all for that matter.

6. It is a useful, but not a necessary condition, of statistical cost allocation that the allocation procedure is carried out at regular intervals. If this condition is fulfilled, the periodic sample size can be reduced.

## THE MECHANICS OF STATISTICAL COST ALLOCATION

We will assume that a book entry exists recording each use of a given resource during a period. The total cost of the resource requires to be allocated between the various users. The record might be an entry in a day-book or a note on a slip of paper. Each entry incorporates a code identifying the user. The total of all of these entries makes up the population from which we are sampling.

A brief resumé of the steps to be employed is as follows:

1. Use random number tables or systematic random sampling to carry out an estimation sampling of attributes on the population. The attribute we are testing is the cost to be charged to each cost centre. From this initial sample, one can estimate the *number* of items in the population which should be charged to each cost centre, but not the value.
2. The *total* cost of the resource to be charged is already known. In step (1) we calculated the *number* of items to be charged, if we assume that *each item has roughly the same value*, this gives us the proportions of the total charge to allocate to each cost centre.
3. We must now test the validity of our allocation by calculating the sampling error in (a) our estimate of proportions and (b) our subpopulation means.

## LEVEL OF CONFIDENCE

The main problem lies in calculating that sample size which will give a prediction within a given sampling error at a given level of confidence. If the sampling error is taken as fixed, say at 5% either side of the predicted value, the size of sample depends on the level of confidence required in the prediction.

For example, with a population of 10,000 items with an estimated mean of £10 and an estimated standard deviation of £2, the sample sizes required to ensure that the actual value lies within approximately 5% of the estimated value at various levels of confidence are:

| Level of confidence | Sample size |
|---|---|
| 99·9% | 170 |
| 99·0% | 106 |
| 95·0% | 62 |
| 90·0% | 44 |

Clearly the level of confidence required is, as usual, a key factor in the economics of statistical cost allocation.

We believe that statistical cost allocation is one of the rare instances in sampling applied to accounting where the chosen level of confidence can be very low, say 80% or even 70%, and yet still achieve the desired result. We justify this opinion as follows.

Where direct cost is used to measure the contribution of a product or process a precise measure is seldom required. The accountant need only calculate whether the contribution is positive or negative to provide a rough measure of the surplus on the job, particularly as this alters through time. *Precise* calculations are, however, seldom required unless the company is working to very narrow profit margins.

The argument is even stronger in the case of cost control. The motivation for allocation here is to prevent the misuse of a resource by charging the user with the direct cost of use. This is a psychological deterrent. Why waste valuable clerical resources in allocating every transaction when the same effect can be achieved by allocating a small sample? The user may be under or overcharged by a small amount in any one period, but, in the long run, these variations will tend to cancel out. The objective of resource conservation can be achieved with considerably less expenditure of clerical effort.

### EXAMPLE

A maintenance and repair department comprising one hundred employees service the plant and equipment in five divisions A to E.

Approximately 10,000 jobs are completed each month ranging from small repair jobs taking less than an hour to sections of major construction projects taking several months. The jobs are divided into two categories, major jobs (estimated budget of over £100) and minor jobs: 95% of the jobs processed are minor jobs. The total value of jobs of £100 or below is £107,605.

A maintenance slip (MS) is filled in for each job completed, detailing division code, nature of job, time and materials consumed, etc. These are forwarded weekly to the costing department where they are listed and summated on a maintenance charge sheet. The totals are charged to individual divisional costing P to L accounts.

The cost accountant decides to release clerical labour by using the statistical cost allocation procedure for handling the *minor* charges which account for 95% of all jobs.

In order to make a first estimate of the proportion of MSs to be allocated to each department a systematic sample of each tenth MS is drawn from the file of 9900 MSs in a given month.

The number, proportion and value of jobs performed in each division is, therefore, estimated to be as follows:

Table 14.1. Estimate of number of jobs performed in each division giving proportion and value.

| Division | No. of jobs performed | Percentage of total | Percentage applied to value (£) |
|---|---|---|---|
| A | 297 | 30% | 32,281 |
| B | 29 | 3% | 3228 |
| C | 148 | 15% | 16,141 |
| D | 75 | 7% | 7532 |
| E | 441 | 45% | 48,423 |
| | 990 | 100% | £107,605 |

The reader will recall that the standard error of a proportion is given by the formula

$$s = \left\{\frac{p(1 - p)}{n}\right\}^{1/2}$$

where $s$ = standard error, $p$ = proportion having condition, $n$ = number of units in the sample.

The standard error and confidence interval at a 90% level of confidence for each proportion is therefore:

Table 14.2. Calculation of confidence limits on proportions.

| (1) Division | (2) Standard error | (3) Confidence interval at 90% level of confidence (a) | (b) |
|---|---|---|---|
| A | 1·10 | 27·6–32·4 | 27·7–32·3 |
| B | 0·55 | 2·1–3·9 | 2·3–3·9 |
| C | 1·13 | 13·2–16·8 | 13·3–16·9 |
| D | 0·81 | 5·7–8·3 | 5·8–8·4 |
| E | 1·58 | 42·4–47·6 | 42·5–47·5 |

Column 3(a) gives the 90% confidence interval calculated by means of the formula given on p. 167. Column 3(b) gives the more precise figure obtained from Arkin (1963), Table F. The difference between the figures is surely not significant on a job of this nature?

A divisional manager is hardly likely to complain if he is *undercharged* for maintenance, therefore the cost accountant can ignore the lower bound of the confidence interval and state to the manager of division A:

'I am charging you 30% of the maintenance charge. I am 95% confident your charge is at least 27·6% of the total charge.'

I doubt whether the second sentence would, in fact, be stated, but it *could* be stated if a manager questioned the maintenance charge.

The reader will have noticed that the standard error is a much greater proportion of small proportions than of large, i.e.

| (a) Proportion | (b) Standard error | (c) $\frac{(b)}{(a)} \times 100$ % |
|---|---|---|
| 45 | 1·58 | 3·5 |
| 30 | 1·10 | 3·6 |
| 15 | 1·13 | 7·5 |
| 7 | 0·81 | 11·6 |
| 3 | 0·55 | 18·3 |

This suggests that we might, perhaps, cut down sample size to estimate the larger proportions, i.e. widen their confidence interval. A moment's thought will show that this action provides no saving. The cost of sampling each of these five populations is a *joint* cost. The sample size is determined by the sample size required to estimate the *smallest* proportion with the required degree of accuracy.

In practise the cost accountant has no idea initially what the smallest proportion will be, so he must take a preliminary sample to estimate this. I suggest a preliminary sample of 200 units as a minimum to achieve this estimate.

The smallest likely proportion determined, the cost accountant can now use the methods described in Chapter 8 to determine sample size.

But, the reader may complain, you have made one very important assumption in the previous calculations which you have not stated explicitly. This assumption is that each maintenance job is of roughly the same value, i.e. £10·87. But the maintenance jobs can vary in value up to £100. Suppose that the jobs done for division B, although only 3% of the total number of jobs were all £100 jobs? And suppose that 40% of jobs done for division E were all £1 jobs. Surely your allocation procedure would give a highly misleading result?

True. But we can test this hypothesis.

The random sample of 990 units that we used to estimate the *proportion* of jobs can also be used to test this hypothesis. We select 49 units from each division and use these to estimate the standard deviation of each section. We then calculate the mean value of the samples. The results come out as follows:

| Division | Mean value (£) | Estimate of SD (£) |
|---|---|---|
| A | 12·7 | 4·2 |
| B | 8·6 | 3·4 |
| C | 9·5 | 6·6 |
| D | 21·2 | 4·4 |
| E | 8·0 | 10·2 |
| Population (actual) | 10·9 | 4·4 |

It could be that the difference between the actual mean population value of £10·9 (£107,605/9900 = £10·9) and the various divisional means is caused by sampling error. However in the case of some divisions, particularly division D, this is highly unlikely. £10·9 stands 2·3 (divisional) standard deviations away from £21·2; only about 2% of the time would this happen by chance according to Table 21.1.

We can make a rough approximation of the reasonableness of these divisional mean values by multiplying them by our estimate of the actual number of units in each of these divisional populations as estimated in Table 14.1. The result is

| Division | Estimate of population size | Mean estimate (£) | Total value (£) |
|---|---|---|---|
| A | 2970 | 12·7 | 37,719 |
| B | 290 | 8·6 | 2494 |
| C | 1480 | 9·5 | 14,060 |
| D | 750 | 21·2 | 15,900 |
| E | 4410 | 8·0 | 35,280 |
| | 9900 | 10·87 | 105,453 |

This total is not too far distant from the actual total of £107,605. The calculation *proves* nothing but it fails to cast doubt on the divisional estimates.

We decide to ensure that these estimates of the divisional mean values are within ±£1 of the *actual* divisional mean value. What sample size do we need to ensure this? The sample sizes are calculated in Table 14.3.

Table 14.3. Calculation of required sample size to ensure that the divisional mean is within ± £1 of actual divisional mean. Confidence level 90%, unit precision limit £1.

| Division | Population | SD | UPL/SD | Required sample size |
|---|---|---|---|---|
| A | 2970 | 4·2 | 0·24 | 47 (297) |
| B | 290 | 3·4 | 0·29 | 40 (29) |
| C | 1480 | 6·6 | 0·15 | 111 (148) |
| D | 750 | 4·4 | 0·23 | 51 (75) |
| E | 4410 | 10·2 | 0·10 | 255 (441) |

The existing sample size, in brackets in Table 14.3, is larger than the required sample size except in the case of division B. In the case of this division we must top up our sample to the required size. Our final estimate of the divisional mean values comes out at:

Table 14.4. Calculation of confidence limits on divisional mean value estimates. (a) Based on con. int. of ± £1. (b) Based on con. int. on total sample.

| Division | Mean value of sample (£) | 90% confidence limits on mean value of population | |
|---|---|---|---|
| | | (a) | (b) |
| A | 12·7 | 11·7–13·7 | 12·3–13·09 |
| B | 8·8 | 7·8–9·8 | 7·92–9·68 |
| C | 9·5 | 8·5–10·5 | 8·62–10·38 |
| D | 21·2 | 20·2–22·2 | 20·34–22·06 |
| E | 8·0 | 7·0–9·0 | 7·20–8·80 |

We can now multiply the estimates of the population size of each division by the estimates of the mean value to arrive at a final estimate.

| Division | Estimate of: population size | mean value (£) | Total £ | Total adjusted £ |
|---|---|---|---|---|
| A | 2970 | 12·7 | 37,719 | 38,468 |
| B | 290 | 8·8 | 2552 | 2604 |
| C | 1480 | 9·5 | 14,060 | 14,339 |
| D | 750 | 21·2 | 15,900 | 16,214 |
| E | 4410 | 8·0 | 35,280 | 35,980 |
| | | | 105,511 | 107,605 |

The total of the final estimates is rather lower than the actual total so we have to shade up all the estimates by a factor of 1·0198 (107,605/105,511 = 1·0198).

How accurate are these estimates? We have calculated confidence limits on the proportions in Table 14.2 and confidence limits on the mean value estimates in Table 14.4. We are for example 90% × 90% = 80% confident that the actual value lies within, approximately, 8% of the estimate in the case of division A.

It is, however, most important to remember that this estimate is being performed every month. As month succeeds month the *absolute* size of the sample increases proportionately.

Therefore, at the end of twelve months in the case of division A we have a sample size of approximately 297 × 12 = 3564 out of a population of approximately 12 × 2970 = 35,640, a huge sample. The standard error on this proportion is around 0·767. Therefore the 90% confidence interval is reduced to between 28·74% and 31·26% and the 90% confidence limits on the mean value estimate falls to £0·12.

Surely these year end estimates are close enough when one considers the rather arbitrary nature of many allocation procedures?

This completes our case study of allocating direct cost to cost centres using statistical sampling. But before we leave this subject let us take a look at several other examples of the use of this method.

## SOME OTHER EXAMPLES

A Xerox copying machine was introduced into a company with several departments. Initially the service was provided free and the cost of duplicating shot through the roof. The Xerox machine was then charged out proportionately to departmental administrative cost. The result was that each department determined to get their fair share of duplicating costs and so these rose even higher. Finally the cost accountant insisted on proper records of usage being kept by the duplicating department. This record took four clerical hours a week to prepare.

Since the sole purpose of the cost allocation procedure was to provide a psychological deterrent against the misuse of the duplicating machine it was an ideal subject for statistical cost allocation. A 10% random sample based on page and line number provided a 90% level of confidence in the allocation of Xerox cost within ±5% for each department. The procedure absorbed one clerical hour as against four for the previous method.

A fleet of chauffeur driven vehicles serviced several departments. The cost was allocated between the departments in proportion to the number of staff in each department permitted to use the service. When asked why the allocation was not based on use, the cost accountant replied that 'the clerical cost would exceed the benefit of having more precise allocation.' Since each chauffeur kept a historical record of every trip including department serviced and mileage, the situation was ideal for the use of statistical cost allocation. Unfortunately, the cost accountant could not be persuaded to accept this advice since he did not understand the principle of statistical sampling.

A transport division shipped surfacing material between a quarry division and a road surfacing division of a given company. The rate of return on the transport division was much less than on quarrying and surfacing. A consultancy firm was employed to improve the efficiency of the transport division. The consultants found that the transport division was in fact as efficient as the three other divisions! However, it acted as a buffer between quarrying and surfacing and was forced to accept costs pushed on to it by other divisions in their efforts to maximize their own performance, i.e. a great deal of lorry capacity was wasted by lorries waiting to load and to tip at surfacing sites.

The consultants suggested that a record of individual lorry utilization should be maintained to enable the cost accountant to credit transport and debit the other two divisions with the lorry capacity they wasted.

The cost accountant did a quick calculation and ridiculed the idea as much too expensive. His argument ran something as follows:

'There are 100 lorries and they make, on the average, five trips a day, that is about 10,000 trips a month! Even if we assume that one of my clerks can process a delivery slip in a minute, which is unlikely, the job would take about 166 hours a month, that is about forty man days a month at the rate my chaps work. At two minutes a slip it would take eighty man days.'

The consultants pointed out that since the procedure would provide around £10,000 a year in additional lorry capacity, it might just be worth it even at an additional 1000 man-days a year of clerical labour. However, this figure could be reduced considerably by using statistical cost allocation methods.

The situation is ideal for using the technique. There is a large population (10,000 a month), none of which are individually significant and absolute precision of allocation is not required. The delivery slips are collected by the cost department during the month and the allocation calculated at the month end.

The problem is to allocate lorry time between the three divisions of quarrying, transport and surfacing. An examination of past records suggest that it is improbable that any division will be charged less than 10% of the total transport cost. 10%, therefore, determines the sample size, as follows:

| | |
|---|---|
| Population | 10,000 |
| Expected proportion | 10% |
| Confidence level | 90% |
| Precision limit on 10% | ±3% |
| Required sample size | 270 |

A sample size of 270 is close to 300 and since Figure 21.2.1 gives us the actual confidence limits on a sample of 300 from a population of 10,000 at a 90% level of confidence we draw a sample of 300.

The sample proportions come out as follows:

| | | Percentage |
|---|---|---|
| Quarrying | 36 | 12 |
| Transport | 195 | 65 |
| Surfacing | 69 | 23 |
| | | 100 |

From Figure 21.2.1 the 90% confidence limits are;

| | Confidence limits % | |
|---|---|---|
| Quarrying | 9 | 16 |
| Transport | 60 | 70 |
| Surfacing | 19 | 27 |

A systematic sample of 300 can be drawn in about two hours and analysed in a further four hours. The 166 clerical hours is reduced to about 6.

Notice that since the allocation is performed each month any excess or underpayment by a department will tend to cancel out in the long run. As month by month the total cumulative sample increases so the estimated proportions approach closer to the actual proportions.

## CONCLUSION

Statistical cost allocation is a viable method of allocating direct cost under certain conditions. The key conditions are that (a) the population of items to be allocated is relatively large, (b) none of these items are individually significant in value and (c) absolute accuracy of allocation is not required.

If the input to the allocation process is entered as a joint cost with some other necessary job to a unit record system, i.e. a computer, it is unlikely that statistical cost allocation will be competitive. Otherwise it is almost invariably the more efficient system.

The method is particularly useful as a cheap psychological deterrent preventing the misuse of resources.

## QUESTION SERIES 14

1. Suggest the ideal conditions for using statistical sampling to allocate direct cost between users of a given facility.
2. 

| | |
|---|---|
| Population | 50,000 vouchers |
| No. of users | 3: A, B and C |
| Total value to be allocated | £245,501 |
| Precision limit required | $\pm 3{\cdot}5\%$ |
| Level of confidence | 90% |

Calculate sample size required to allocate direct cost between users. The minimum proportion is estimated to be 10% and the maximum 50%. (*)

3. The mean value and proportion of each strata in the sample drawn in problem (2) above is;

| User | Percentage | Mean value | Estimated standard deviation |
|---|---|---|---|
| A | 48 | 4·65 | 2·1 |
| B | 41 | 4·98 | 1·8 |
| C | 11 | 5·75 | 4·1 |

Do these figures suggest that we should reject the hypothesis that the mean value of individual jobs for all three users are the same? (*)

4. Suggest a use for statistical cost or revenue allocation from your own experience.

## SOME ANSWERS TO QUESTION SERIES 14

2. Since the precision limits are fixed at 3·5% the largest percentage determines sample size.

50% = 500 see Figure 21.2.1
10% = 300 see Figure 21.2.1

Therefore sample of 500 is drawn.

3.

| Stratum | Estimated population size | Sample size | Mean | 90% confidence limit |
|---|---|---|---|---|
| A | 24000 | 240 | 4·65 | 4·44–4·86 |
| B | 20500 | 205 | 4·98 | 4·77–5·12 |
| C | 5500 | 55 | 5·75 | 4·84–6·66 |

The population mean is 4·91. It is rather improbable that the true mean of stratum A is 4·91.

# 15

# Can the Bayesian Approach Assist the Auditor?

## INTRODUCTION

The statistical method used in this book up to date is often called the 'classical' method. The term 'classical' implies that the inference about the population is based entirely on the information derived from the sample. The inference is 'objective' in the sense that the prior beliefs, prejudices, intuitions, etc., of the auditor have not been allowed to colour the auditor's inference about the population.

Some critics of this rather austere approach claim that by ignoring the auditor's intuition and experience in assessing the information from the sample we are throwing out the baby with the bathwater. The auditor's intuition and past experience ought to be used to *improve* the objective inference from the sample.

It is argued that the subjective beliefs of the auditor should be wedded to the objective information from the sample to provide the final assessment of the accounting population under audit.

A branch of statistics called Bayesian statistics has been used in other branches of business activity to wed subjective and objective information into a unified inference. Therefore it has been suggested by writers such as Kraft (1968) and Tracy (1969) that the Bayesian approach might be useful to auditors.

Bayesian statistics is a subject of much controversy in the statistical world and I am not competent to express a view on the *theoretical* viability of the method.

This chapter will attempt to explain the basic Bayesian approach to auditing and examine some of the arguments put forward by the advocates of the method.

## SUBJECTIVE PROBABILITY

Suppose that two political parties, A and B, are fighting an election. The legislature has 1000 seats and so the winning party must gain at least 501 seats.

Before the election you are asked 'Which party will win the election?'. You say 'Party A'. However the questioner presses you harder by asking, 'How many seats do you think Party A will win?'. You say 600. The tenacious questioner goes even further and asks 'How confident are you that Party A will win 600 seats?'.

You could answer 'I am very confident', or 'I think they have a reasonable chance' or 'I am not really very confident'.

These statements are *subjective estimates* of the probability of Party A winning the election. The probability is expressed in words rather than numbers but it is no less of a probability estimate for that.

There is, however, nothing to prevent you expressing your final estimate in the form of a table of numbers such as Table 15.1.

Table 15.1. Probability of Party A winning *n* seats at next general election.

| My estimate of number of seats to be won by Party A | Probability |
|---|---|
| 0–399 | 0·00 |
| 400–499 | 0·20 |
| 500–599 | 0·50 |
| 600–699 | 0·25 |
| 700–1000 | 0·05 |
| | 1·00 |

Table 15.1 provides a more comprehensive picture of your subjective estimate. You think Party A will win yet you have allocated 20% of the total probability to Party B winning.

If a political journalist were canvassing a cross-section of the electorate on the question of whom they think will win the next general election, an arithmetic statement such as Table 15.1 would provide a more precise assessment than a qualified opinion.

It is clear that the same principle could be applied to auditing. *Before* an auditor tests an accounting population he can ask himself 'What is the probability that the error rate will exceed a given proportion?'. His answer can be tabulated as in Table 15.2.

This subjective estimate is based upon the auditor's prior experience with this and other populations, his intuition and any other supplementary knowledge he possesses.

Table 15.2. Auditor's subjective estimate of error rate in population *before* he draws sample from population.

| Possible error rate in population | Auditors subjective probability estimate of this percentage occurring |
|---|---|
| 0·00–0·001 | 0·10 |
| 0·001–0·01 | 0·50 |
| 0·01–1·00 | 0·30 |
| 1·00–2·00 | 0·10 |
| 2·00–3·00 | 0·00 |
| 3·00–100·00 | 0·00 |
| | 1·00 |

## OBJECTIVE PROBABILITY

In all of the examples prior to this chapter we have used objective probabilities. By 'objective' we mean that the probabilities depend upon the evidence from the sample drawn from the population.

We measure or estimate the *parameters* of the population, mean, standard deviation, etc., and calculate the probability of some condition arising assuming that the population has these characteristics.

For example in Chapter 3 p. 15 we provided a table illustrating the probabilities of 0, 1, 2, 3, . . . , $n\%$ of a sample being defective if 10% of the population from which the sample was drawn was defective.

Table 15.3. If sample of 100 contains two errors the table provides the objective probability of population containing $n\%$ of erroneous units. By $n\%$ we mean all possible error rates above $(n-1)\%$ to $n\%$, i.e. 3% includes all error rates above 2%.

| (a) Possible error rate in population | (b) Objective probability of population containing this percentage of errors if sample contains 2% of error |
|---|---|
| 1 | 0·190 |
| 2 | 0·271 |
| 3 | 0·222 |
| 4 | 0·146 |
| 5 | 0·085 |
| 6 | 0·045 |
| 7 | 0·023 |
| 8 | 0·011 |
| 9–100 | 0·007 |
| | 1·000 |

Conversely we can calculate the objective probabilities of a population being $x\%$ defective if $y\%$ of a sample drawn from that population is defective.

For example if we find two errors in a sample of 100 drawn from a very large population the probabilities of the population containing various proportions of error is shown in Table 15.3. The most probable proportion of error in the population is 2% and the table shows that we can be 90% certain that the number of errors in the population does not exceed 5%.

These probabilities are objective in the sense that their truth follows as a logical deduction from our assumptions about the size and shape of the population.

The advocates of the Bayesian approach claim that by merging his *a priori* subjective probabilities with the objective probabilities inferred from the sample the auditor can arrive at a better overall assessment than he could by using the objective probabilities alone.

Let us work through an example using the Bayesian approach.

## TESTING A DOUBTFUL DEBTS PROVISION

A company is owed a large number, say 10,000, of small debts. Since none of these debts are individually significant the doubtful debts provision is based on the *proportion* of the year end debts which are considered to be doubtful. In past years this proportion has always fallen between 2% and 6% with a mean of 3%. The proportion is always estimated to the nearest 1%. The company have set up a doubtful debts provision based on 5%. Is this adequate?

First the auditor must set down his subjective estimate of the probability of various doubtful debt proportions occurring in the population. Using his past experience as a guide he makes the following estimates.

| Percentage of doubtful debts | Probability of this percentage occurring |
|---|---|
| 1 | 0·10 |
| 2 | 0·20 |
| 3 | 0·40 |
| 4 | 0·15 |
| 5 | 0·10 |
| 6 | 0·05 |
| 7–100 | 0·00 |
| | 1·00 |

Next he decides to draw and test a random sample of debts to see what proportion of these are doubtful debts. *He uses a lower confidence level and wider precision limit than he would if he were using objective probability on its own.*

The figures are as follows:

| | |
|---|---|
| 1. Population | 10,000 |
| 2. Level of confidence | 90% |
| 3. Precision limit | ±3% |
| 4. Estimated percentage | 3% |

The required sample size is 150 units. The auditor draws a random sample of 150 units and finds that 5% of these are doubtful debts. Now 5% lies within the confidence interval 3% ± 3% but it is at the upper end of the interval. Under normal circumstances the auditor would probably extend his sample size to reduce his precision limit, i.e.

| Precision limit | Sample size |
|---|---|
| ± | |
| 3 | 150 |
| 2 | 260 |
| 1 | 770 |

That is he will adopt a new precision limit, of say, 3% ± 2% and extend his sample to 260 units.

However, the Bayesian theorists argue that there is an alternative strategy available. The auditor can use his *a priori* subjective probability estimates to improve his level of confidence.

But before we return to subjective probability let us calculate the precise objective probability of the proportion of doubtful debts being 1%, 2%, 3%, . . . , 10% from the sample information now before us.

We drew a sample of 100 accounts from the population and discovered that 5% of the sample are doubtful accounts. If 5% of a sample of 100 have a given condition the most likely proportion in the total population is 5%, but the population proportion *could* be anywhere between 0·5% to 99·5%.

The actual objective probabilities are set out in Table 15.4. From Table 15.4 we see that the probability of the population proportion being exactly 5% (actually between 4% and 5%) is only 0·176 or 17·6%. However there is a 91% probability that the population proportion lies between 2% and 9%.[1]

Let us now examine Table 15.5 when we compare the pre-sample subjective probabilities in column (b) against the post-sample objective probability in column (c).

Table 15.4. If sample has 5% of units with given condition what are the objective probabilities that this sample is drawn from population with 1, 2, 3, . . ., *n*% with given condition? Note: 5% includes all percentages above 4% up to and including 5%.

| (a) Possible percentage of doubtful debts in population | (b) Probability of this sample being drawn from population listed in (a) if 5% of sample are doubtful debts |
|---|---|
| 1 | 0·003 |
| 2 | 0·036 |
| 3 | 0·101 |
| 4 | 0·156 |
| 5 | 0·176 |
| 6 | 0·161 |
| 7 | 0·128 |
| 8 | 0·091 |
| 9 | 0·061 |
| 10–100 | 0·087 |
| | 1·000 |

The objective probabilities are more pessimistic than the subjective. The former cluster around 5% while the latter cluster around 3%. Also the scatter of possibilities is much wider in the case of the objective probabilities.

We now come to the key question in this chapter. Is it meaningful for the auditor to combine the subjective with the objective probabilities to arrive at a composite probability figure which he can use to decide whether or not a doubtful debts provision based on 5% of accounts is satisfactory?

Table 15.5 provides the necessary calculations.

The subjective probabilities of various percentages of doubtful debts is multiplied by the objective probabilities to arrive at the composite figure displayed in column (d). Since the total of column (d) is less than unity

[1] These calculations are rather approximate. The true 90% confidence interval is 2%–10·3%.

the various figures in column (d) must be scaled up by 100/96 to make the total probability equal to one. This has been done in column (e). Column (f) provides the final composite cumulative joint probability of the doubtful debt proportion not exceeding various percentages of the population.

Does column (f) provide the auditor with a better measure of evaluation than the simple objective probability table provided in Table 15.4?

Table 15.5. Calculation of cumulative joint probability % of doubtful debt % in the population being less than a given percentage.

| (a) Possible percentage of doubtful debts % | (b) Subjective probability | (c) Objective probability from sample | (d) Joint probability | (e) Adjusted joint probability | (f) Cumulative joint probability % |
|---|---|---|---|---|---|
| 1 | 0·10 | 0·003 | 0·000 | 0 | 0 |
| 2 | 0·20 | 0·036 | 0·007 | 0·07 | 7 |
| 3 | 0·40 | 0·101 | 0·040 | 0·42 | 49 |
| 4 | 0·15 | 0·156 | 0·023 | 0·24 | 73 |
| 5 | 0·10 | 0·176 | 0·018 | 0·19 | 92 |
| 6 | 0·05 | 0·161 | 0·008 | 0·08 | 100 |
| 7 | 0 | 0·128 | 0 | 0 | 100 |
| 8 | 0 | 0·091 | 0 | 0 | 100 |
| 9 | 0 | 0·061 | 0 | 0 | 100 |
| 10–100 | 0 | 0·087 | 0 | 0 | 100 |
| | 1·00 | 1·00 | 0·096 | 1·00 | |

The classical approach assumes that any state of the population is as likely to occur as any other state. In other words if the error rate can be 1%, 2%, 3%, . . ., 100%, to the nearest 1%, then each of these states has a probability of 0·01 of occurring since each of the 100 states is equally likely to be present.

The Bayesian approach claims that this assumption of equality of all possible states is neither likely nor desirable.

'There is nothing inherently "right" or "scientific" in assuming that all states of nature are equally likely of being true. The best that can be said for this assumption is that it is neutral. But if there is strong evidence against a neutral assumption, the auditor should turn to the Bayesian approach.'[1]

The Bayesian statistician attaches different weights to the various possible states. He determined these weights by using his prior knowledge, experience and intuition.

[1] Tracy (1969).

Thus in the example given in Table 15.5 the Bayesian statistician will argue that column (f) provides a more accurate inference than column (c) since column (c) ignores the prior experience and intuition of the auditor.

Column (c) suggests that there is only about a 50% chance that the doubtful debt proportion is 5% or below. Column (f) suggests that the probability is as high as 92%!

Which statistic we believe depends upon the weight we are prepared to attach to the auditor's subjective judgement.

## USING KRAFT'S TABLES TO CALCULATE LEVEL OF CONFIDENCE

W. H. Kraft (1968) has provided a set of tables which can assist the auditor to arrive quickly at a cumulative joint probability.

The table illustrated on p. 184 is for a sample size of 200.

Column (a) provides three possible cumulative subjective probabilities each less optimistic than the previous one. Column (b) gives various likely error rates up to 5%.

The auditor will use the table as follows.

Suppose his subjective estimate of the error rate is the third grouping, i.e. the most pessimistic. He draws a random sample of 200 units and finds three errors.

He looks down the column headed '3' and finds a series of numbers opposite the various likely error rates, i.e.

| Error rate | Confidence level |
|---|---|
| | % |
| 0·001 | 0·2 |
| 0·01 | 72·2 |
| 0·02 | 95·6 |
| 0·03 | 99·2 |

This tells him that if he finds three errors in a sample of 200, *given his prior subjective probabilities*, he can be 95·6% confident that the error rate in the population does not exceed 2%.

The table saves the auditor the chore of calculating these probabilities *ab initio* each time, as was done earlier in this chapter.

Kraft (1968) also provides tables for sample sizes of 150 and 100.

## WHAT HAS THE BAYESIAN APPROACH ACHIEVED?

The purpose of applying statistical methods to auditing is to attach numbers to evaluations which were previously verbal. I believe that this makes auditing more scientific by allowing comparisons of evaluation to be made between different years and between different audits and auditors.

The classical statistical method permits the auditor to use his experience and intuition as guides to selecting the required level of confidence, precision limits, and criteria for stratification. The Bayesian method goes further by allowing the auditor to attach numbers to his prior beliefs about the *likely* state of the population he is auditing.

Table 15.6. Table for calculating cumulative joint probability (CJP) from given sample size. Sample size = 200. (a) Optimistic (prior belief). (b) Average. (c) Pessimistic.

| Cumulative Prior probability | Possible error rates ($r$) | Number of errors in sample | | | | | |
|---|---|---|---|---|---|---|---|
| | | 0 | 1 | 2 | 3 | 4 | 5 |
| 0·60 | 0·001 | 922 | 535 | 098 | 010 | 001 | 000 |
| 0·90 | 0·01 | 998 | 977 | 912 | 811 | 651 | 437 |
| 0·95 | 0·02 | 1000 | 997 | 985 | 955 | 885 | 753 |
| (a) 0·98 | 0·03 | 1000 | 1000 | 999 | 995 | 982 | 948 |
| 0·99 | 0·04 | 1000 | 1000 | 1000 | 999 | 995 | 985 |
| 1·00 | 0·05 | 1000 | 1000 | 1000 | 1000 | 1000 | 1000 |
| 0·25 | 0·001 | 697 | 185 | 022 | 002 | 000 | 000 |
| 0·90 | 0·01 | 997 | 981 | 953 | 902 | 801 | 627 |
| 0·95 | 0·02 | 1000 | 998 | 992 | 977 | 935 | 836 |
| (b) 0·98 | 0.03 | 1000 | 1000 | 999 | 997 | 990 | 965 |
| 0·99 | 0·04 | 1000 | 1000 | 1000 | 999 | 997 | 990 |
| 1·00 | 0·05 | 1000 | 1000 | 1000 | 1000 | 1000 | 1000 |
| 0·25 | 0·001 | 744 | 218 | 025 | 002 | 000 | 000 |
| 0·75 | 0·01 | 990 | 937 | 850 | 722 | 542 | 340 |
| 0·90 | 0·02 | 1000 | 996 | 984 | 956 | 894 | 782 |
| (c) 0·95 | 0·03 | 1000 | 1000 | 998 | 992 | 975 | 934 |
| 0·98 | 0·04 | 1000 | 1000 | 1000 | 999 | 995 | 986 |
| 1·00 | 0·05 | 1000 | 1000 | 1000 | 1000 | 1000 | 1000 |



I believe that this fact in itself, is a useful addition to auditing technique, since it forces the auditor to clarify his thoughts on what he considers to be normal.

When the information from the objective sample becomes available his prior subjective estimate will help him to evaluate the information from the objective sample. Thus the method is of value even if the auditor declines to perform the final step of multiplying the two probabilities together to arrive at a final cumulative joint probability (CJP).

The CJP from a single audit is of little value, but if the auditor, over time, builds up a fund of experience in the use of CJPs this figure will provide a numerical representation of his intuitive feeling about the outcome of an audit.

As such, the concept of the CJP could make a major contribution to the science of auditing. As Oskar Morgenstern has said:

'Even though a residuum always remains that cannot be quantified, it can frequently be made smaller. It is through constant expansion of measurement where there was none, better measurement where it was poor, that we extend our knowledge. Where we have measurement we do not need intuition to determine the characteristics of what was measured. Our intuitive efforts can then be better concentrated on the areas that still remain inaccessible for measurement.'

To date our experience in the use of the CJP is very limited. We can only progress by trial and error. We must be grateful to the advocates of the Bayesian method for suggesting a novel and interesting approach to an old problem.

In 1962 the American Institute of Certified Public Accountants set up a committee to look into the contribution of statistical sampling to auditing.

The kernel of their report is contained in the following passage:

'Statistical sampling appears to be most useful in dealing with numerous items where the purposes of the test are not intertwined with those of other tests. For the present, at least, its greatest usefulness likely relates to audit tests having essentially a single objective. In areas where the independent auditor is dealing with items related to other areas of the audit in such a way that a conclusion is not drawn mainly upon the basis of the single test, meaningful application of statistical sampling in any comprehensive sense may present considerable difficulties. . . . Further experimentation may, however, uncover ways of dealing with such interrelationships. . . .

'No techniques have as yet been brought to the attention of the committee whereby the qualitative audit satisfaction based on such reviews and procedures can be combined mathematically or statistically with the quantitative information gained from detailed audit tests.'[1]

Even if Bayesian statistics does not answer this problem it surely takes the auditor a long stride in the right direction?

## QUESTION SERIES 15

1. What is meant by the classical method in statistics?
2. How does the Bayesian approach modify the classical approach to making an inference about a population from a sample? (*)

[1] Committee on Statistical Sampling (1962).

3. Set out a table giving your subjective estimates of the probability of the Labour Party winning the following number of seats at the next general election in the UK.

0–100
101–200
201–300
301–400
401–500
501–600.

Assume that 315 seats gives an overall majority.

4. How does a subjective probability estimate differ from an objective probability estimate? (*)

5.

| Possible value (£) | Probability Subjective | Objective |
|---|---|---|
| under 500 | 0·05 | 0·15 |
| 500–599 | 0·20 | 0·40 |
| 600–699 | 0·45 | 0·25 |
| 700–799 | 0·25 | 0·15 |
| over 800 | 0·05 | 0·05 |
| | 1·00 | 1·00 |

(a) Calculate the cumulative joint probability of the value exceeding the given figures.
(b) What is the probability that the value will not exceed £799? (1) classical (2) Bayesian. (*)

6. A sample of 200 is drawn from a population. Use Kraft's table on p. 184 to calculate the cumulative joint probability under the following conditions.

| No. of errors | Population error rate does not exceed | Subjective estimate |
|---|---|---|
| 1 | 0·01 | Pessimistic |
| 2 | 0·03 | Optimistic |
| 3 | 0·01 | Average |
| 4 | 0·02 | Pessimistic |
| 5 | 0·01 | Optimistic |

(*)

7. Even if the auditor does not combine his objective and subjective probabilities into a single probability what use might the subjective estimate be?

## SOME ANSWERS TO QUESTION SERIES 15

2. Bayesian approach assumes that all states of the variable are not equally likely to occur.
4. All of the information on which an objective inference is based is derived from a sample drawn from the population. A subjective estimate is based on many factors derived from experience, intuition, etc.
5. (a)

| Possible value (£) | Joint probability | Adjusted JP | Cumulative JP |
|---|---|---|---|
| Under 500 | 0·007 | 0·029 | 0·029 |
| 500–599 | 0·080 | 0·333 | 0·362 |
| 600–699 | 0·113 | 0·471 | 0·833 |
| 700–799 | 0·037 | 0·154 | 0·987 |
| Over 800 | 0·003 | 0·013 | 1·000 |
| | 0·240 | 1·000 | |

(b) (1) 95%
(2) almost 99%

6. (1) 93·7% (2) 99·9% (3) 90·2% (4) 89·4% (5) 43·7%

# 16

# Some Advice on Integrating Statistical Sampling into Conventional Auditing Procedures

## INTRODUCTION[1]

Many professional accounting firms have set out with great enthusiasm to apply statistical sampling methods to auditing. The majority of these attempts have come to grief. In this article I will describe some of the pitfalls that lie in the path of the unwary auditor.

Statistical sampling differs in many respects from conventional sampling. Instructions must be more precise and supervision more strict. A glossary of new words unfamiliar to most accountants needs to be learned and the staff at all levels must learn several new concepts.

Social psychologists have pointed out that the most effective way of preventing an organization from adopting a new idea is to 'innoculate' the organization against it by applying it *unsuccessfully* at a low level in the organization. Much too often statistical sampling has been introduced by a single partly trained enthusiast without the informed support of the top management of his firm. He invariably picks the wrong sort of application to start off with and uses imperfectly trained clerks without proper documentation to carry out the audit. The result is a disaster which plays into the hands of the conservative element in the firm who resent any form of change (and which accounting firm does not have an entrenched conservative element?).

I have come across this pattern of events in spots as far apart as Glasgow, Scotland, Cork, Ireland and Pretoria, South Africa.

The solution is to provide adequate time, thought and resources to the problem of integrating scientific sampling procedures into conventional audit practice. The following pages will provide some advice, culled from

[1] An earlier version of this chapter appeared in *The Australian Accountant*, **41** 5, June 1971, 202–207.

the author's experience in setting up scientific sampling procedures in several professional offices. I do not claim that this list provides either a necessary or sufficient list of dos and don'ts to guarantee the successful implementation of scientific sampling; it ought, however, to guide the reader away from the more obvious pitfalls.

### 1. ENSURE INFORMED SUPPORT FROM THE TOP

It often happens that the idea of using scientific sampling in auditing is introduced below the partnership level by a manager or senior auditor or even by a student accountant.

If the method is to have any chance of success it must first be sold to top management, in this case to the partnership level.

I suggest that a course lasting at least one day, and preferably two, be held to introduce the partners to the basic concepts of economic sample size, confidence levels, precision limits, etc., and to give the partners a chance to work examples themselves. With sampling confidence comes by doing. Most of the arguments against statistical sampling arise from a misunderstanding of the method. These difficulties can be most easily cleared up within the context of a practical example. My experience suggests that *if* a partners' course is to be held then *only* partners should be invited to attend it. Expertise in tax, consolidations, etc., does not guarantee a rapid assimilation of sampling theory—and the main benefit of the course comes from the question sessions.

Once the basic theory of scientific sampling is understood and accepted a formal decision should be taken at partnership level to the effect that the firm accepts the validity of scientific sampling and intends to test it out over a given period on a given job.

In my experience it is better to get every partner to commit himself to statistical sampling before the experiment begins. This avoids any later wrangling at high level which can upset the sampling team.

It is also advisable to agree on a minimum period for the test of, say, three years. The first year is always difficult and expensive since the firm is at the bottom of a learning curve. *Three years is the minimum time period within which the validity of scientific sampling can be judged.* Unless a firm is prepared to commit itself for this period of time it is not worth while starting the experiment.

In summary, before a professional firm begins to use scientific sampling methods in its auditing procedures a formal *informed* decision on this matter should be unanimously voted at partnership level. I also suggest that the validity of scientific sampling cannot be properly tested in a period of less than three years.

## 2. ENSURE THAT INFORMED OPINION ABOUT SCIENTIFIC SAMPLING IS PROVIDED AT EVERY LEVEL OF THE FIRM

I do not believe that scientific sampling can be applied successfully within a professional firm unless one of the partners has a good grasp of the basics of sampling theory. A good many problems of principle will crop up in the early days of applying the technique. These will need to be discussed and explained at partnership level. The case for scientific sampling needs to be argued lucidly and forcibly. In my experience this cannot be done effectively by an employee or by an outsider. The momentum of the early enthusiasm (which soon evaporates) can only be maintained by a partner who is committed to the method. There is no need for this partner to be expert in the mathematics of sampling but he must be quite clear on the concepts of confidence level, precision limits, dispersion, the economics of sample size and so on.

The *expert* on scientific sampling may be a partner but more usually he will be a manager. Probably a university graduate who has done some statistics as part of his degree. He will rarely be a fully qualified statistician.

The expert may be brought in from another firm, but since such experts are scarce they are also expensive. It is cheaper, and probably more efficient in the long run, to train an insider. Psychological tests are available for selecting individuals with mathematical ability, but in the writer's experience the experts select themselves. I suggest that the following order of time span is needed to allow an individual to become trained in scientific sampling.

1. Two day attendance at an introductory course run by one of the professional institutes.
2. Something of the order of fifty hours of systematic private learning.

In one example known to the author a qualified accountant with no previous sampling knowledge was given a fortnight of free time to learn scientific sampling from two books and twelve selected articles.[1] At the end of this period he was tested by myself and he had obtained a sound grasp of the *basic elements* of sampling theory as applied to auditing. He did, however, have access to myself (by phone) during the fortnight

[1] The books were Arkin (1963) and Vanasse (1968). The articles included Smurthwaite (1965), Hall (1967), Davidson (1959), Haworth (1969), Vance (1960) and Trentin (1968).

to clear up difficulties as they arose.[1] This does not suggest that the cost of 'learning time presents an insuperable barrier to applying statistical sampling'.

The diffusion of knowledge of scientific sampling methods below the managerial level presents no problem. Most of the professional examinations now include questions on sampling and inference and most student accountants find scientific sampling methods a good deal more interesting than traditional sampling methods. The problem is not 'how can I persuade these young chaps to use scientific sampling' but rather 'how can I stop these young chaps applying scientific sampling to jobs where the preliminary analysis for scientific sampling has not yet been done'. Once the basic concepts are introduced student accountants soon *teach themselves* about scientific sampling by arguing points out on the job. At least this has been my experience.

The educational problem lies at the top not the bottom of the professional firm.

## 3. MUST WE HAVE ACCESS TO A FULLY QUALIFIED STATISTICIAN WHEN DESIGNING THE SAMPLING PLAN?

If a statistician is available he should obviously be consulted. But remember that *he* will have to learn about auditing and accounting practice.

Since audit sampling concerns large and not small sample theory (the former is a good deal easier to handle than the latter), and since accounting populations are stable and relatively easily defined, the statistical theory is not difficult. The auditor can always increase his sample size if he is in doubt. The advantage of the qualified statistician is that he will be able to provide an answer at a given level of confidence with a smaller sample size, or at a higher level of confidence with the same sample size.

The answer to this question must be pragmatic. Of course we would prefer the assistance of a statistician trained in accounting practice but such paragons are scarce, and if we have to wait until they become generally available scientific sampling will take years to get off the ground.

However, as a rider I would add that if you can get a statistician to review your sampling plan in the *first year* of the audit, this should prove helpful.

We should remember that physicists, chemists, psychologists, marketing analysts, etc., all use standard statistical techniques in their work as a matter of course without consulting trained statisticians.

[1] He phoned me six times during the fortnight and raised about twenty queries.

## 4. HOW DO WE SELECT THE FIRST APPLICATION?

The novel concepts and techniques involved in scientific sampling present problems enough without compounding the difficulties by selecting a difficult initial application.

What are the characteristics we should look for in an initial application?

First we should look at the size of the population. There can be no doubt that scientific sampling becomes more economical the larger the population to be audited.[1] The benefits become particularly evident when the population size exceeds a figure of the order of 10,000. Let us then select our initial application from a job with a population in excess of 10,000 units.

A second useful characteristic is *homogeneity*, i.e. that the units in the population to be audited are not too dissimilar from one another. Alternatively, if the population to be audited, say debtors, includes a few large units, let us ensure that we are able to stratify the population by segregating out these large units into a separate subpopulation (see Figure 16.1).

A non-homogeneous population will have a relatively large standard deviation and a relatively large sample size.

A third useful characteristic is to pick a population which is operationally viable. By this I mean that the sampling can be carried out relatively easily.

The following are characteristics you want to avoid *first time out*.

1. Populations which are not numbered or numbered out of sequence.
2. Populations which are geographically dispersed or maintained by different branches using slightly different systems.
3. Populations which are likely not to be complete.
4. Untidy, ill-maintained populations of accounting records, such as small-time hire purchase accounts maintained by part time clerical labour.

Finally choose as your initial experiment in statistical sampling an account which you know well and about which you have a good deal of prior information.

I have come across several cases of a firm deciding to use scientific sampling for the first time on a new job because conventional methods are

[1] This results from the fact that the level of confidence in the inference from a sample depends upon the absolute size of the sample rather than upon the relation between the sample size and the population size.

not suitable for one reason or another. This is a great mistake as the reader will quickly find out if he tries it!

I would emphasize that scientific sampling can be, and has been, successfully applied to every one of the examples given above, but it is not wise to tackle these types of jobs as *a first experimental application.*

The above advice concerns the characteristics of the population to be audited; it tells us nothing of the best sampling technique to start on. Should we start with attribute or variables sampling, i.e. estimating proportions or values? If we select attribute sampling should we use estimation sampling of attributes (estimating a proportion), or its simplified versions acceptance or discovery sampling (estimating if a proportion is above or below a given percentage).

This is a difficult question to answer. Let us dispose of acceptance sampling first. This is a tricky method to apply and should be avoided as a first application. Discovery sampling is very easy to understand and apply but it simply tells us that the error rate is less that $x\%$ with a given level of confidence.[1] Estimation sampling of attributes provides more information by estimating the range within which the actual error rate lies; this method is also relatively easy to apply.

However, there can be little doubt that the big pay-off in using scientific sampling comes from variables sampling, such as estimating the value of inventory, bad debts, etc. It is a pity that so many writers have suggested that the main contribution of scientific sampling to auditing is in estimating error proportions. *This is quite definitely not so.*

Since the big pay-off in using scientific sampling comes from estimating *values* rather than proportions, I believe that novitiates should take their courage in both hands and select variables sampling as their first experimental application, preferably using the MUS system set out in Chapter 17.

I put forward this recommendation for three reasons.

1. The inferences from variables sampling are so much more clear cut than the inferences from estimating error proportions. The auditor states 'I am 95% confident that the value of debts six months overdue lies between £40,000 and £50,000.' Rather than 'I am 95% confident that the proportion of errors is less that 3%.'
2. A wider precision (confidence) interval can usually be accepted in variables sampling. This reduces the required sample size.
3. Although variables sampling is a little more complicated than attributes sampling, the difference is not substantial.

[1] Discovery sampling also tends to reject rather a large number of 'good' batches with the 'bad' batches.

For these reasons I favour variables sampling using the MUS method as a first application. Variables sampling provides a useful and demonstrable end result, attribute sampling of error proportions simply ensures that auditors do not check uneconomically large sample sizes.

## 5. WHEN YOU APPLY SCIENTIFIC SAMPLING TO AN AUDIT FOR THE FIRST TIME PREPARE A JOB MANUAL, DETAILED AUDIT PROGRAMME AND PROCEDURE BOOKLET

Since scientific samples tend to be smaller than conventional audit samples, audit clerks must follow instructions very precisely. Both the selection and audit of the sample needs to be carried through with a rigour which, in my experience, is foreign to conventional audit practice.

This means that in the first year of applying scientific sampling to an audit a good deal of preliminary work needs to be done.

The first stage is for a decision to be taken at partnership level about the *level of confidence* and *precision limits* to be applied. The confidence level used under normal circumstances is likely to be 90%. This can be dropped to 80% if the consequences of being wrong are not too serious. It can be raised to 99% on important parts of the audit. I suggest that the procedure should be adopted that audit managers normally use 90% and must provide specific reasons for varying this.

The *precision limit* (confidence interval) tells us the acceptable sampling error, i.e. the probable error of inference caused by the sample not including the entire population. I believe that it is of the nature of auditing that the precision limit should be very narrow, say no more than ±2%, except in rare circumstances. One example where a wider precision limit may be viable is in estimating the proportion of debts $x$ months overdue. The precision limit can be extended to, say, ±5%.

If the confidence level is pushed above 95% and/or the precision limit reduced below ±2% the required sample size begins to rise rather dramatically. Table 4.2 illustrates the sample size required at various confidence levels and precision limits. The cost of being wrong would have to be very high to induce an auditor to select a confidence level of 99% and a precision limit of ±1% or less!

The confidence level and precision limit having been decided upon the next step is to prepare the *scientific sampling job manual.* Table 16.1 sets out the section headings in my own manual. Notice that the manual is divided into three sections.

1. Characteristics of population to be audited
   (a) this year (b) last year

2. Characteristics of sample
   (a) this year (b) last year
3. Characteristics of auditor.

It is advisable for the auditor to state clearly the inference he has drawn from the sample *and to commit himself by signing this inference.*

In addition to the job manual, which is self-explanatory, a *scientific sampling audit programme* should be prepared in the first year. This should set out in complete detail the various steps the auditor will take when (a) calculating standard deviation (if required) and sample size, (b) drawing the sample, and (c) making inferences from the sample.

Table 16.1. Main headings in scientific sampling job manual.

| | | 1973 | 1972 |
|---|---|---|---|
| *Auditing firm*: | ABC | | |
| *Client*: | XYZ Ltd. Address | | |
| *Dept. or division* | Credit control sector, accounting dept. | | |
| *Details re population to be audited* | | 1973 | 1972 |
| 1. Job | | | |
| 2. Condition sought | | | |
| 3. Period covered | | | |
| 4. From no. to no. | | | |
| 5. Estimated size of population | | | |
| 6. Estimated mean of population | | | |
| 7. Estimated standard deviation of population | | | |
| 8. Estimated skewness of population | | | |
| 9. Pre-sampling estimate of value or proportion having condition | | | |
| 10. Type of sampling to be employed | | | |
| 11. Statistical tables used | | | |
| *Details re sample* | | | |
| 1. Level of confidence required | | | |
| 2. Precision limit adopted | | | |
| 3. Advance estimate of standard deviation | | | |
| 4. Required sample size | | | |
| 5. Proportion having condition or mean value of sample | | | |
| 6. Vouchers missing | | | |
| 7. Inference from sample | | | |
| | signed: | | |
| *Details re auditor* | | | |
| 1. Name | | | |
| 2. Position | | | |
| 3. Period of audit | | | |
| 4. Special comments re audit | | | |

I suggest that a *standard procedure booklet* be prepared which covers such questions as 'What should the auditor do if:

(a) A voucher is missing?
(b) The standard deviation estimate differs significantly from last year?

(c) The file sequence has been changed?

(d) The value and, therefore, importance of the job has changed significantly since last year?

(e) Sample size very much larger this year than last?

etc.

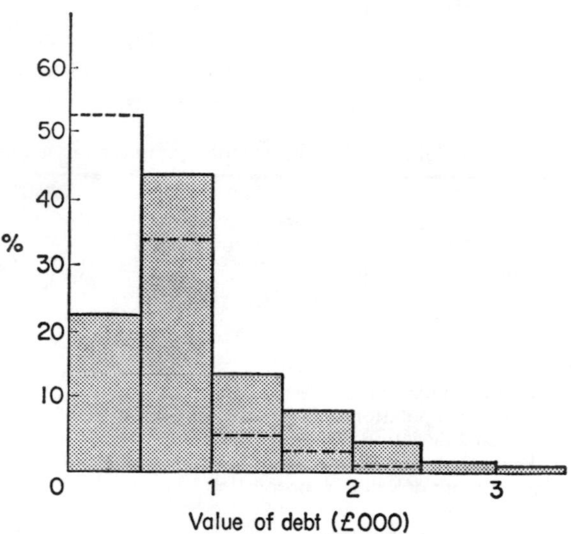


Fig. 16.1. Population of debts stratified into six classes, illustrating skewness. Notice that the population is skewed out towards the right, but the skewness is not too severe.

| Value of debt lies between £ | No. of debts in this class | Total value of this class £ | Percentage of total |
|---|---|---|---|
| 0–499 | 1278 (54%) | 319,500 | 23 |
| 500–999 | 835 (35%) | 626,250 | 46 |
| 1000–1499 | 164 (7%) | 205,000 | 15 |
| 1500–1999 | 73 (3%) | 127,750 | 9 |
| 2000–2499 | 26 (1%) | 58,500 | 4 |
| 2500–2999 | 8 (—) | 22,000 | 2 |
| over 3000 | 2 (—) | 10,000 | 1 |
| | 2386(100%) | 1,369,000 | 100 |

The procedure booklet should also cover such topics as:

(a) Accepting help from the client.
(b) Statistical tables available.
(c) Procedure for selecting random numbers.

etc.

In addition to all this, in the first year I would advise the auditor to carry out a careful analysis of the population to be audited. In particular a reasonably accurate measure of the standard deviation and skewness of the population is useful.[1] A frequency distribution such as that shown in Figure 16.1 is helpful for measuring skewness.[2]

Since accounting populations are very stable through time this analysis need not be repeated every year. A stratification every four or five years is sufficient unless the company being audited is a fast growing company.

## 6. OTHER POINTS TO CONSIDER

Should the auditor inform his client that he is switching to scientific sampling? I think he should. It shows the client that he is keeping abreast of current developments in auditing and, if the client's employees are informed, it can provide a useful psychological deterrent to carelessness or fraud. We may recall that when the US Inland Review service first announced that it was handing over the analysis of taxpayers' income tax returns to a computer, hundreds of worried tax-dodgers wrote in admitting prior underdeclarations of income. So many in fact that the fines went a long way towards paying for the computer!

Another reason in favour of informing the client is that he might notice that the sample sizes drop significantly and think the auditor is skimping the job!

Although the first experimental job should be handled by a team of specialists I would strongly advise against handing all statistical sampling audits to a specialist team. Once scientific sampling is adopted by a firm it should become a technique to be used on *any* suitable audit by *any* member of the auditing staff.

If the client's records are on a computer, the auditor can use the computer to assist in the scientific sampling procedures.[3] For example the computer can be programmed to:

[1] Unless the MUS system is being used.

[2] If the population is stored on a computer tape or disc, this analysis should present no problems.

[3] See Irvine (1964) in selective bibliography and Chapter 11.

1. Select the random samples.
2. Stratify the population as in Table 16.1.
3. Test the viability of systematic or cluster sampling.
etc.

I see little danger in using the client's computer staff to programme these operations. The auditor will, of course, provide the string of random numbers.

## 7. THE NEED FOR A LIBRARY

I have lectured from time to time on the subject of applying scientific sampling to audit procedures. I find that the same questions tend to be asked again and again. Nearly all of these questions have been carefully answered in one or another of the many articles written on statistical sampling. But how seldom does one find an auditor, even an auditor who has used scientific sampling methods for some time, who has read a fraction of the available literature! I strongly recommend that every professional firm using scientific sampling methods builds up a library of books and articles on the subject.

As a starter I append a selective bibliography in Chapter 20. This includes all the books and a short selection of some of the more interesting and pertinent articles.

## SUMMARY

If a professional firm decides to integrate scientific sampling methods into its conventional auditing procedures it should take note of the following points:

1. Ensure that *informed* support for the method exists at the partnership level, and that informed opinion is accessible at every level although a fully qualified statistician need not be employed.
2. Take great care in selecting the first application. Ensure that the firm commits itself to the method for a long enough period to give the method a fair chance of success.
3. Prepare proper documentation and a detailed audit programme.
4. Use the client's computer to assist in the audit if one is available.
5. Build up a library of books and articles on statistical sampling—and *read them.*

Statistical sampling is a proven method of improving audit practice but unless a firm is prepared to commit an adequate amount of time and resources to *learning* the basic concepts and *integrating* the method into its procedures I would strongly advise against the firm attempting to use it.

## QUESTION SERIES 16

1. If you were asked to design a one day course on 'Statistical Sampling in Auditing', what are the four key ideas you would try to put across? (*)
2. What is the minimum period which must elapse before one can judge whether or not statistical sampling is a success or failure?
3. Suggest a scheme for training an auditor in scientific sampling.
4. Roughly how many hours must be spent to acquire a sound grasp of the basic elements of statistical sampling?
5. If a qualified statistician is available what use would you make of him in designing a statistical sampling plan?
6. The first time you use statistical sampling on an audit you should avoid jobs with certain characteristics. Name four of these.
7. What statistical sampling technique should an auditor use in his first application? Give reasons.
8. Set out instructions for work required to be done in *first year* of audit using statistical sampling.
9. Set out headings in statistical sampling job manual.
10. Design a statistical sampling audit programme for any job set out in this book.

## SOME ANSWERS TO QUESTION SERIES 16

1. (a) Accuracy of inference from sample related to absolute size of sample rather than to fraction of population in sample.
   (b) Statistical sampling provides a *precise* measure of auditor's confidence in outcome of tests.
   (c) Random sampling removes all degree of bias in selecting sample.
   (d) Auditor's judgement not supplanted but used to select confidence level required and to stratify population.

# 17

# Monetary Unit Sampling

## INTRODUCTION

In 1961, Haskins and Sells, one of the leading firms of accountants in the United States, decided to examine the applicability of statistical sampling techniques to their audit procedures. The task of reviewing existing procedures was deputed to Kenneth Stringer, a senior executive of the firm.

After a lengthy review of existing methods, Stringer concluded that the statistical sampling procedures available at that time were not suited to the context of auditing. He described various limitations in the existing procedures (Stringer, 1963) and set out to devise a scientific system of audit sampling which would be rigorous, in the sense that the conclusions from the audit could be evaluated in numerical terms, and yet operationally simple enough to be applied by audit clerks not trained in statistical theory.

In devising this system, Stringer called in the assistance of Professor Frederick Stephan, a statistician, of the University of Yale. Professor Stephan was a member of the New York Business Advisory Committee and an enthusiastic advocate of the application of statistical theory to solving business problems.

The original system devised by Stringer and Stephan in 1962 has been modified in several aspects during the intervening years. Also, our research indicates that the method of application is not standardized in all countries. The method described below is our own modification of the original system.

## A DESCRIPTION OF THE MUS SYSTEM

As with all scientific sampling systems the basic approach is to calculate a required sample size, select a random sample of this size and estimate the value of error in the population from the sample. The monetary unit sampling (MUS) system differs from other systems previously described in that, although it attempts to measure error value, neither the number of units in the population nor the standard deviation of the population need be known.

The system requires (i) that the total value of the population be known and (ii) that the auditor must decide on a given level of confidence. This latter is translated into what is termed a *reliability factor* in the MUS system. Finally, the auditor must state his requirement for *monetary precision.* This latter term is roughly equivalent to the concept of precision limits in other systems.

In summary the required data for calculating sample size using the MUS system is:

1. Total value of population, *V*.
2. Reliability factor, *R*.
3. Monetary precision, *P*.

We will now take a more detailed look at the method of calculating the reliability factor and the monetary precision.

## THE METHOD OF DECIDING ON THE RELIABILITY FACTOR

The confidence level chosen by the auditor about his estimate of the amount of error in the population depends upon his assessment of the likely effectiveness of the internal control system. If he concludes that the internal control system is very effective, he will require a relatively low level of confidence in the inference drawn from his sample. That is to say he will be prepared to place some degree of confidence in his prior belief in the effectiveness of the internal control system in reducing error. If the internal control is weak or non-existent, then the auditor must place his entire confidence in the inference from the sample. The distribution of his level of confidence between prior belief in the effectiveness of the system and inference from the sample is illustrated in Figure 17.1. The horizontal axis measures the auditor's declining belief in the effectiveness of the internal control system. The vertical axis measures the auditor's level of confidence in the inference from the sample. The shaded triangular area FGE can be regarded as the auditor's confidence in the effectiveness of the internal control system. The area FABG can be taken to represent the degree of risk acceptable to the auditor.

The reader will recognize the similarity between this type of approach and the Bayesian approach to auditing advocated by such writers as

Kraft (1968) and Tracy (1969). The MUS system builds an overall confidence level by combining the prior subjective belief of the auditor with the objective information from the sample.

The MUS system, at a later stage, tests the validity of the auditors prior belief about the effectiveness of the internal control system by testing for compliance errors in the sample drawn. If the sample invalidates the auditors prior belief in the effectiveness of the system the precision limit is enlarged by an appropriate amount. This adjustment will be explained later.

Table 17.1 is provided as part of the MUS system. We note that levels of confidence ranging between 50% and 99% are available to the auditor. The confidence level selected depending upon his prior assessment of the

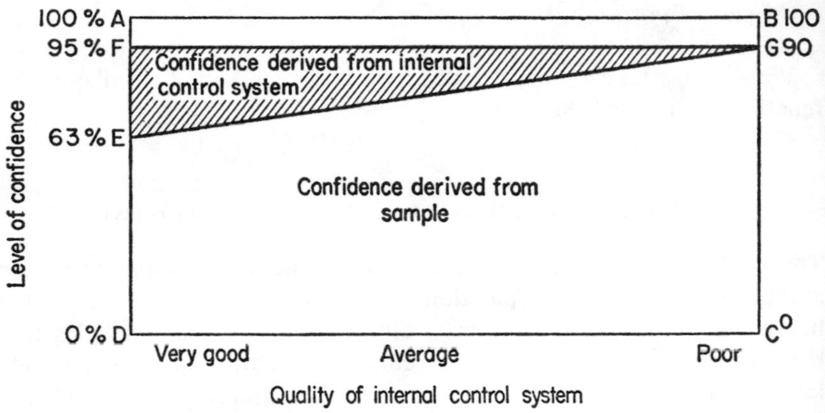


Fig. 17.1. Diagram to illustrate make-up of auditor's confidence in his assessment of maximum likely error value in population. The overall confidence is made up partly from sample information and partly from a prior estimate of effectiveness of internal control system in eliminating error.

effectiveness of the internal control system. The confidence level is then translated into what the MUS system calls a *reliability factor*. This factor is generated by the cumulative Poisson distribution at zero sampling error. The reliability factor from Table 17.1 is coded as $R$ in the MUS system.

Our research suggests that the following confidence levels and reliability factors are commonly associated with the given evaluation of the internal control system.

| Evaluation of internal control system | Confidence level | Reliability factor |
|---|---|---|
| Very good | 63% | 1 |
| Average | 86% | 2 |
| Rather poor | 95% | 3 |

The statement on auditing standards issued by the AICPA auditing standards executive committee provided a table for which the combined reliability level desired is assumed to be 95% (AICPA, 1973, p. 53).

| Auditors judgement concerning reliance to be assigned to internal accounting control and other relevant factors % | Resulting confidence level for substantive tests % |
|---|---|
| 90 | 50 |
| 70 | 83 |
| 50 | 90 |
| 30 | 93 |

## THE METHOD OF DECIDING ON MONETARY PRECISION

In every audit the auditor ought to decide on a minimum unacceptable error value (MUEV) for the population as a whole. This is the minimum overall value of error which is *significant* in the sense that if this value is exceeded, a much more rigorous audit will be required. If it is exceeded by a large amount, say double the MP, then the auditor will refuse to accept the population being audited.

The MUS system forces the auditor to place a specific value on the MUEV of error in the population. This amount is entitled the *monetary precision* (MP) required by the auditor.

The MP on any given audit is usually decided by the partner responsible for the audit. The amount of the MP is fixed in relation to the value of the entire audit and *not* related to the individual populations making up the audit.

Our research suggests that the MP is often fixed at 1% of sales or 2% of net asset value, although these rule of thumb measures are not suited to all audits.

Since the MP is fixed in relation to the total value of the audit it can be a very large fraction of individual populations within the total audit. Thus very small samples will be called for from these subpopulations.

Table 17.1. Table for converting confidence level to reliability factor and associated precision adjustment factor.

| Reliability Factors($R$) | 0·7 | 0·8 | 0·9 | 1·0 | 1·1 | 1·2 | 1·3 | 1·4 | 1·6 | 2·0 | 2·3 | 3·0 | 4·6 |
|---|---|---|---|---|---|---|---|---|---|---|---|---|---|
| Confidence Levels | 50% | 55% | 59% | 63% | 66% | 69% | 72% | 75% | 80% | 86% | 90% | 95% | 99% |
| | Precision adjustment factors($p$) for evaluating samples at confidence levels shown above | | | | | | | | | | | | |
| Rank of errors[a] | For errors of overstatement of population items[b] | | | | | | | | | | | | |
| 1 | 1·01 | 1·05 | 1·11 | 1·15 | 1·20 | 1·24 | 1·28 | 1·32 | 1·39 | 1·51 | 1·59 | 1·75 | 2·04 |
| 2 | 1·01 | 1·04 | 1·08 | 1·12 | 1·15 | 1·18 | 1·21 | 1·24 | 1·28 | 1·38 | 1·44 | 1·56 | 1·77 |
| 3 | 1·00 | 1·04 | 1·07 | 1·10 | 1·13 | 1·15 | 1·17 | 1·20 | 1·24 | 1·31 | 1·36 | 1·46 | 1·64 |
| 4 | 1·00 | 1·03 | 1·06 | 1·09 | 1·11 | 1·13 | 1·15 | 1·17 | 1·21 | 1·27 | 1·32 | 1·40 | 1·56 |
| 5 | 1·00 | 1·03 | 1·05 | 1·08 | 1·10 | 1·12 | 1·14 | 1·16 | 1·19 | 1·25 | 1·29 | 1·36 | 1·50 |
| 6 | 1·00 | 1·03 | 1·05 | 1·07 | 1·09 | 1·11 | 1·13 | 1·14 | 1·17 | 1·23 | 1·26 | 1·33 | 1·46 |
| 7 | 1·00 | 1·02 | 1·04 | 1·07 | 1·09 | 1·10 | 1·12 | 1·13 | 1·16 | 1·21 | 1·24 | 1·31 | 1·43 |
| 8 | 1·00 | 1·02 | 1·04 | 1·06 | 1·08 | 1·10 | 1·11 | 1·12 | 1·15 | 1·20 | 1·23 | 1·29 | 1·40 |
| 9 | 1·00 | 1·02 | 1·04 | 1·06 | 1·08 | 1·09 | 1·10 | 1·12 | 1·14 | 1·19 | 1·22 | 1·28 | 1·38 |
| 10 | 1·00 | 1·02 | 1·04 | 1·06 | 1·07 | 1·09 | 1·10 | 1·11 | 1·14 | 1·18 | 1·21 | 1·26 | 1·36 |
| 11 | 1·00 | 1·02 | 1·04 | 1·05 | 1·07 | 1·08 | 1·10 | 1·11 | 1·13 | 1·17 | 1·20 | 1·25 | 1·35 |
| 12 | 1·00 | 1·02 | 1·03 | 1·05 | 1·07 | 1·08 | 1·09 | 1·10 | 1·13 | 1·16 | 1·19 | 1·24 | 1·34 |
| 13 | 1·00 | 1·02 | 1·03 | 1·05 | 1·06 | 1·08 | 1·09 | 1·10 | 1·12 | 1·16 | 1·18 | 1·23 | 1·33 |
| 14 | 1·00 | 1·02 | 1·03 | 1·05 | 1·06 | 1·07 | 1·08 | 1·10 | 1·12 | 1·15 | 1·18 | 1·22 | 1·32 |
| 15–19 | 1·00 | 1·02 | 1·03 | 1·05 | 1·06 | 1·07 | 1·08 | 1·09 | 1·11 | 1·15 | 1·17 | 1·22 | 1·31 |
| 20–24 | 1·00 | 1·01 | 1·03 | 1·04 | 1·05 | 1·06 | 1·07 | 1·08 | 1·10 | 1·13 | 1·15 | 1·19 | 1·26 |
| 25–29 | 1·00 | 1·01 | 1·03 | 1·04 | 1·05 | 1·06 | 1·06 | 1·07 | 1·09 | 1·11 | 1·13 | 1·17 | 1·24 |
| 30–39 | 1·00 | 1·01 | 1·02 | 1·03 | 1·04 | 1·05 | 1·06 | 1·06 | 1·08 | 1·10 | 1·12 | 1·15 | 1·22 |
| 40–49 | 1·00 | 1·01 | 1·02 | 1·03 | 1·04 | 1·05 | 1·05 | 1·06 | 1·07 | 1·09 | 1·10 | 1·13 | 1·19 |
| 50–74 | 1·00 | 1·01 | 1·02 | 1·03 | 1·04 | 1·04 | 1·05 | 1·05 | 1·06 | 1·08 | 1·09 | 1·12 | 1·17 |
| 75–99 | 1·00 | 1·01 | 1·02 | 1·02 | 1·03 | 1·03 | 1·04 | 1·04 | 1·05 | 1·07 | 1·08 | 1·10 | 1·14 |
| 100 and over | 1·00 | 1·01 | 1·02 | 1·02 | 1·03 | 1·03 | 1·04 | 1·04 | 1·04 | 1·06 | 1·07 | 1·09 | 1·12 |

| Rank of errors[a] | For errors of understatement of population items[b] | | | | | | | | | | | | |
|---|---|---|---|---|---|---|---|---|---|---|---|---|---|
| 1 | 0·67 | 0·58 | 0·49 | 0·45 | 0·40 | 0·35 | 0·31 | 0·28 | 0·22 | 0·14 | 0·10 | 0·05 | 0·00 |
| 2 | 0·96 | 0·92 | 0·87 | 0·82 | 0·78 | 0·74 | 0·70 | 0·66 | 0·60 | 0·49 | 0·42 | 0·30 | 0·14 |
| 3 | 0·99 | 0·95 | 0·91 | 0·88 | 0·84 | 0·81 | 0·78 | 0·76 | 0·71 | 0·62 | 0·57 | 0·46 | 0·29 |
| 4 | 0·99 | 0·95 | 0·92 | 0·90 | 0·87 | 0·85 | 0·82 | 0·80 | 0·76 | 0·69 | 0·64 | 0·54 | 0·39 |
| 5 | 0·99 | 0·96 | 0·93 | 0·91 | 0·89 | 0·87 | 0·85 | 0·83 | 0·79 | 0·73 | 0·68 | 0·60 | 0·45 |
| 6 | 0·99 | 0·96 | 0·94 | 0·92 | 0·90 | 0·88 | 0·86 | 0·84 | 0·81 | 0·75 | 0·71 | 0·64 | 0·51 |
| 7 | 0·99 | 0·97 | 0·95 | 0·93 | 0·91 | 0·89 | 0·87 | 0·86 | 0·83 | 0·77 | 0·74 | 0·67 | 0·54 |
| 8 | 0·99 | 0·97 | 0·95 | 0·93 | 0·91 | 0·90 | 0·88 | 0·87 | 0·84 | 0·79 | 0·76 | 0·69 | 0·57 |
| 9 | 0·99 | 0·97 | 0·95 | 0·94 | 0·92 | 0·90 | 0·89 | 0·88 | 0·85 | 0·80 | 0·77 | 0·71 | 0·60 |
| 10 | 0·99 | 0·97 | 0·96 | 0·94 | 0·92 | 0·91 | 0·90 | 0·88 | 0·86 | 0·81 | 0·78 | 0·73 | 0·62 |
| 11 | 1·00 | 0·98 | 0·96 | 0·94 | 0·93 | 0·91 | 0·90 | 0·89 | 0·86 | 0·82 | 0·79 | 0·74 | 0·64 |
| 12 | 1·00 | 0·98 | 0·96 | 0·94 | 0·93 | 0·92 | 0·90 | 0·89 | 0·87 | 0·83 | 0·80 | 0·75 | 0·65 |
| 13 | 1·00 | 0·98 | 0·96 | 0·95 | 0·93 | 0·92 | 0·91 | 0·90 | 0·87 | 0·84 | 0·81 | 0·76 | 0·66 |
| 14 | 1·00 | 0·98 | 0·96 | 0·95 | 0·94 | 0·92 | 0·91 | 0·90 | 0·88 | 0·84 | 0·82 | 0·77 | 0·67 |
| 15–19 | 1·00 | 0·98 | 0·96 | 0·95 | 0·94 | 0·93 | 0·91 | 0·90 | 0·88 | 0·85 | 0·83 | 0·78 | 0·68 |
| 20–24 | 1·00 | 0·98 | 0·96 | 0·96 | 0·95 | 0·94 | 0·93 | 0·92 | 0·90 | 0·87 | 0·85 | 0·81 | 0·73 |
| 25–29 | 1·00 | 0·98 | 0·97 | 0·96 | 0·95 | 0·94 | 0·93 | 0·93 | 0·91 | 0·89 | 0·87 | 0·83 | 0·76 |
| 30–39 | 1·00 | 0·98 | 0·97 | 0·96 | 0·96 | 0·95 | 0·94 | 0·93 | 0·92 | 0·90 | 0·88 | 0·85 | 0·78 |
| 40–49 | 1·00 | 0·99 | 0·98 | 0·97 | 0·96 | 0·95 | 0·95 | 0·94 | 0·93 | 0·91 | 0·90 | 0·87 | 0·80 |
| 50–74 | 1·00 | 0·99 | 0·98 | 0·97 | 0·96 | 0·96 | 0·95 | 0·95 | 0·94 | 0·92 | 0·91 | 0·88 | 0·83 |
| 75–99 | 1·00 | 0·99 | 0·98 | 0·98 | 0·97 | 0·97 | 0·96 | 0·96 | 0·95 | 0·93 | 0·92 | 0·90 | 0·86 |
| 100 and over | 1·00 | 0·99 | 0·98 | 0·98 | 0·97 | 0·97 | 0·96 | 0·96 | 0·96 | 0·94 | 0·93 | 0·91 | 0·88 |

[a] This column refers to the rank of extrapolated errors ($E_0$). Errors of overstatement and of understatement should be ranked separately, and within each group the ranking should be from the largest to the smallest amount of error.

[b] The distinction between errors of overstatement and of understatement should be based on their effect on the population from which the sample was drawn. If this is a reciprocal population to the one of primary audit interest, as in sampling subsequent invoices in search for understatement of recorded liabilities, errors will have an opposite effect on the population of primary audit interest.

This table is taken from the Auditape System Manual which was developed by Haskins and Sells, the American certified public accounting firm.

## CALCULATING THE SAMPLE SIZE

Once the reliability factor, $R$, and the monetary precision, $P$, have been decided upon, the calculation of the sample size is a relatively simple operation.

The next step is to divide the MP by the reliability factor $R$. The result of this division is described as the '$J$ factor'. This is the sampling interval expressed in pound sterling.

The required sample size is found by dividing the total value of the population to be audited, £$V$, by this $J$ factor. Thus

$$\text{Sample size} = \frac{V}{J}$$

Once the sample size has been calculated, exponents of the MUS system have experimented with a variety of methods for drawing the sample. All of the methods used have attempted to relate the probability of selection of an item from a population to the value of that item. By this means they have attempted to neutralize the extreme skewness inherent in most accounting populations. One method is described by Meikle (1972) but the most popular method which we will now describe is based on Anderson and Teitlebaum (1973).

'The modification incorporated in dollar unit sampling is conceptually simple. The trick is merely to define the sampling unit in the population not as being an individual invoice or an individual receivable balance, but as being an individual dollar.'[1]

The calculation of the $J$ factor has been described above, we note that this factor is expressed as so many units of money value. Under the monetary unit system of sampling every $J$th pound or dollar is selected for audit testing from the population of £$V$. 'Of course, when he selects an individual (pound) he does not verify that individual (pound) by itself. Rather it acts as a hook and drags a whole account balance with it.'[2] Thus the probability that a given account will be selected is roughly proportionate to its value.

An example of monetary unit sampling is provided in Table 17.2.

As the comptometer operator sums the inventory column, she selects out the first account using a random start between £1 and £356, that is £97. She then selects on the values.

[1] Anderson and Teitlebaum (1973), p. 34.
[2] Anderson and Teitlebaum (1973), p. 35.

| | £ |
|---|---|
| 97 | 97 |
| 97 + 356 = | 463 |
| 463 + 356 = | 819 |
| 819 + 356 = | 1175 |
| 1175 + 356 = | 1531 |

Therefore the accounts 1, 4, 8 and 10 are selected for audit. The random start between 1 and 356 can be selected from a random number table or by using some random number generating device such as the junior digits on a banknote.

**Table 17.2.** Illustration of selecting sample using monetary unit sampling. *J* factor, £356, random start, £97.

| Account no. | Account value (£) | Cumulative amount | £ selected | Account selected |
|---|---|---|---|---|
| 1 | 134 | 134 | 97 | 1 |
| 2 | 18 | 152 | | |
| 3 | 93 | 245 | | |
| 4 | 257 | 502 | 463 | 4 |
| 5 | 155 | 657 | | |
| 6 | 8 | 665 | | |
| 7 | 54 | 719 | | |
| 8 | 634 | 1353 | 819, 1175 | 8 |
| 9 | 126 | 1479 | | |
| 10 | 95 | 1574 | 1531 | 10 |

The monetary unit sampling method has certain characteristics which should be noted at this stage.

1. Any item of value equal to or in excess of £$J$ is automatically selected.
2. Any item which is two or more times greater than £$J$ is selected several times. Naturally it is only audited once.
3. The system assumes that the maximum amount by which an item may be in error is the amount of the item. This is true for items which are *overstated*,[1] but not true of items which are *understated.* There is no limit to possible understatement.
4. The probability that a particular item will be selected is roughly proportionate to its value.
5. If a single item in the population is made up of a number of smaller items the MU sampling system can be extended to subsample from this secondary population.

[1] Unless there is a negative balance.

### TYPE OF ERROR

The MUS system differentiates between two types of error. *Monetary error*, where the value stated in the account is incorrect, and *compliance error*, where the value stated is correct but some procedural rule has been violated.

An example of the former type of error is the overstatement of a debt. An example of the latter type is a cheque being signed by one authorized signatory when it ought to have been signed by two.

The MUS system handles monetary and compliance error in different ways. Monetary errors affect the value attributed to monetary precision directly, compliance errors affect the MP only indirectly.

We will explain these adjustments later in this chapter. But first let us work through an example using the MUS system. In this case we will assume that *no errors are found.*

### EXAMPLE

| | |
|---|---|
| Total value of all populations audited | £5,000,000 |
| Internal control system | about average |
| Confidence level selected | 90% ($R$) |
| Monetary precision required | £50,000 ($P$) |

The various stages are as follows:

1. Convert confidence level to reliability factor by using Table 17.1.

   90% converts to 2·3

2. Compute the $J$ factor.

   $$J = P/R = 50{,}000/2{\cdot}3 = £21{,}000$$

3. Compute sample size.

   $$\text{Sample size} = £V/J = 5{,}000{,}000/21{,}000 = 240$$

4. Select random starting point between £1 and £21,000 from beginning of population to be audited. Let us assume that this is £13,826.
5. Using the 13,826 pound as the starting point, select a sample of 240 in the following manner.

   (1) 13,826
   (2) 13,826 + 21,000
   (3) 13,826 + (2 × 21,000)
   (4) 13,826 + (3 × 21,000)
   to
   (240) 13,826 + (239 × 21,000)

Select out those 240 accounts within which these individual pounds happen to fall. This may result in a sample of less than 240.[1]

6. Audit these 240[1] items for monetary and compliance error. If *no errors are found* the proponents of the MUS system claim that we now have a 90% confidence in the statement that the total value of error in the population is less than £50,000. The reader will recognize the similarity between this approach and the method called discovery sampling.

## THE DISCOVERY OF MONETARY ERROR

The system described above has the virtue of great simplicity. The system is simple so long as no errors are discovered. But what happens when a monetary or compliance error is discovered?

Let us first deal with monetary error.

## ADJUSTING THE MP FOR MONETARY ERROR

1. If a monetary error is discovered the next step is to calculate the *estimated value* of the error.

If the item in error is of £$A$ and the value of the error is £$E$, where $E \leqslant A$, the estimated value is calculated by using the formula

$$E \times J/A = £X.$$

If, for example, an account of £800 is in error by £200[2] and the $J$ factor is £6000, the estimated value of the error is taken to be $200 \times 6000/800 =$ £$1500 = X$.

In other words the whole $J$ block containing the error is assumed to be tainted by error in the same proportion as the item sampled.

2. Where several monetary errors are discovered the errors are segregated in the following way.

First errors of overstatement are separated from errors of understatement.

Second those errors contained in items equal to or exceeding $J$ in value are segregated. These are called top stratum errors. Third the remaining errors of overstatement and errors of understatement are *ranked* in order

[1] If some items are two or more times $J$ the sample will be less than 240.
[2] The £800 is said to be 25% tainted.

of *estimated* value. The higher the estimated value the higher the rank of the error.

3. Each estimated value of error is now multiplied by a factor taken from Table 17.1 on p. 204. Each column in Table 17.1 is linked to a given level of confidence shown at the head of the column. To select the requisite precision adjustment factor, the auditor must decide
(1) Whether the error is an overstatement or an understatement.
(2) The rank of the error among all of the errors of that category.
4. The total of the adjusted understatements are now deducted from the total of the adjusted overstatements. The difference being added to the MP.

Let us take an example.

*R* factor 2·3 (90% confidence)
*J* factor £500

| (a) Error value (*E*) | (b) *J* factor | (c) Item value | (d) Estimated value of error |
|---|---|---|---|
| Errors of overstatement | | | |
| (1) 626 | 500 | 637 | 626 |
| (2) 345 | 500 | 400 | 431 |
| (3) 90 | 500 | 360 | 125 |
| 1061 | | | |
| Errors of understatement | | | |
| (4) 680 | 500 | 927 | 680 |
| (5) 275 | 500 | 325 | 423 |
| (6) 100 | 500 | 100 | 500 |
| 1055 | | | |

Column (d) is calculated by multiplying column (a) by the factor (b) divided by (c). This does not apply if column (c) equals or exceeds column (b). In this latter case no *estimate* of error is required, the value of column (a) being written straight into column (d).

The next step is to bring in a conservative adjustment on our estimated errors, that is on all of those errors contained in items of value less than *J*, items (2), (3), (5), (6) in the example. To make this adjustment we use Table 17.1 on p. 204. First we segregate over from undervalued errors.

Second we rank each of these two groups in order of estimated value. Third we read off the appropriate precision adjustment factor from Table 17.1. These are listed under the given level of confidence column, 90% in this case.

The calculations are as follows:

| | Estimated value | Precision adjustment factor (from Table 17.1) | | Adjusted estimated value (£) |
|---|---|---|---|---|
| Overstatement | | | | |
| | 626 | —[a] | | 626 |
| (1) | 431 | 1·59 | | 685 |
| (2) | 125 | 1·44 | | 180 |
| | | | | 1491 |
| Understatement | | | | |
| | 680 | —[a] | 680 | |
| (1) | 500 | 0·10 | 50 | |
| (2) | 423 | 0·42 | 178 | 908 |
| Adjustment to MP | | | | 583 |

[a] No adjustment required since this is the *actual* not estimated error value.

If no errors had been discovered the auditor would have been 90% confident that the total value of error in the population was less than $R \times J = 2{\cdot}3 \times 500 =$ £1150. But monetary error was discovered, so, if the confidence level remains at 90% the precision limit (MP) must be increased to compensate. The new MP becomes 1150 + 583 − 7 = £1726. The £7 is the net value of the actual errors discovered and corrected (1061 − 1055).

## ADJUSTING FOR COMPLIANCE ERROR

If the value of items affected by compliance error is significant, and we return to the definition of significant later, the MUS system will adjust the reliability factor upwards by a given amount. In this way, the system adjusts for the fact that the sample indicates that the auditor's initial assessment of the effectiveness of the internal control system has been optimistic.

For example if the initial confidence level is 63% suggesting a prior belief that the internal control system is a very good one, and the sample suggests otherwise, then the auditor may decide to raise his confidence level to 90%. This will raise the reliability factor from 1·0 to 2·3. This in turn will create an increased monetary precision (MP). This new MP may or may not be acceptable. A larger sample may have to be drawn. The MUS system assumes that the drawing of a second sample will be a rare event.

The formal procedure for moving from one reliability factor to another is as follows:

1. Write down each item affected by a compliance error. These are all assumed to be overstatements.
2. Calculate expected value of error as above.

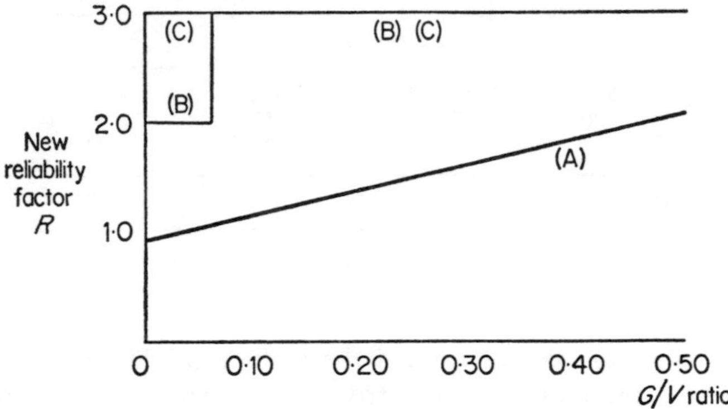


Fig. 17.2. The above diagram allows the auditor to test the reliability of the initial confidence level selected (reliability factor). The $G/V$ ratio on the horizontal axis will select a given reliability factor on the vertical axis. Whether one selects lines A, B or C depends upon the initial assessment of the internal control system. A, Original control system good. B, Original control system average. C, Original control system poor.

3. Multiply each item, in order of estimated value, by the relevant precision adjustment factor using the adjustment factors from the 95% level of confidence column in Table 17.1. This is the level of confidence used for zero rating for effectiveness of internal control.
4. Add the results of these calculations together. Call the result £*F*.
5. Add *F* to three times[1] the *J* factor. Call the result *G*.

[1] The reliability factor for a 95% level of confidence is three.

6. Calculate the ratio $G/V$.
7. Go to Figure 17.2 to select a new reliability factor $R$ depending on the result of the ratio $G/V$.

Let us take an example.

### EXAMPLE

In a given audit the following initial conditions apply.

1. Value $V$, of population £1,500,000.
2. Monetary precision, $P$, required £12,000.
3. The internal control system is thought to be 'good' so an initial confidence level is set at 69% with a reliability factor of 1·2.
4. The $J$ factor is 12,000/1·2 = £10,000.

A sample of 150 items is drawn and the following items contain significant compliance errors.

| Item | Value of item £ |
|---|---|
| (1) | 1000 |
| (2) | 5000 |
| (3) | 2000 |

Does this value of compliance error suggest that the initial reliability factor of 1·2 should be increased?

The procedure to follow to answer this question is as follows:

1. The amounts in error are

| | £ |
|---|---|
| (1) | 5000 |
| (2) | 2000 |
| (3) | 1000 |

2. Since the entire item is in error and no item exceeds $J$, the expected values of all items are equal to the $J$ factor.

$$
\begin{aligned}
5000 \times 10{,}000/5000 &= 10{,}000 \\
2000 \times 10{,}000/2000 &= 10{,}000 \\
1000 \times 10{,}000/1000 &= 10{,}000
\end{aligned}
$$

3. We now multiply each item by the precision adjustment factor from the 95% confidence column (overstatement) of Table 17.1.

(1) 10,000 × 1·75 = 17,500
(2) 10,000 × 1·56 = 15,600
(3) 10,000 × 1·46 = 14,600
$F$ = 47,700

4. Add $F$ to three times the $J$ factor.

$$47{,}700 + (3 \times 10{,}000) = 77{,}700 = G$$

5. Divide $G$ by $V$ the value of the population.

$$77{,}700/1{,}500{,}000 = 0{\cdot}052 = G/V$$

Since the initial internal control system was evaluated as 'good' we read off the new reliability factor from section A of Figure 17.2 as 1·1. This is lower than the initial reliability factor of 1·2 so no adjustment to the MP is required. The sample has confirmed the auditors assessment of the effectiveness of the internal control system.

If the $G/V$ ratio generates a new reliability factor *above* the initial factor a new MP must be calculated.

If in the previous example the $G/V$ ratio had turned out to be, say, 0·153, the $R$ factor would be increased from 1·2 to 1·3. Thus the new MP would be calculated as JR = MP or 10,000 × 1·3 = 13,000. Any monetary error would have to be added to this.

### A WORKED EXAMPLE OF THE MUS SYSTEM

Population value ($V$) = £10,000,000
Monetary precision ($P$) = £64,000
Internal control system = average
Confidence level required ($C$) = 86%

The steps are as follows:

1. Convert confidence level to reliability factor using Table 17.1.

$$86\% = 2{\cdot}00 = C$$

2. Compute the $J$ factor.

$$J = P/C$$
$$J = 64{,}000/2 = £32{,}000$$

3. Compute sample size ($S$).

$$S = V/J = 10{,}000{,}000/32{,}000 = 312 \text{ items}$$

4. A random point within the first £32,000 is selected, say £23,514. The item containing the 23,514th pound is selected, followed by

$$\begin{aligned}
£23{,}514 + 32{,}000 &= £55{,}514\\
£23{,}514 + (2 \times 32{,}000) &= £87{,}514\\
\vdots \quad \vdots \quad\quad \vdots \quad &\;\vdots \quad\; \vdots\\
£23{,}514 + (311 \times 32{,}000) &= £9{,}975{,}514
\end{aligned}$$

5. These 312 items are tested for monetary or compliance error. The following monetary errors are found.

| | Value of item | Value of error |
|---|---|---|
| Errors of overstatement | | |
| (1) | 30,326 | 2135 |
| (2) | 17,524 | 6827 |
| (3) | 2756 | 1000 |
| | | 9962 |
| Errors of understatement | | |
| (1) | 47,200 | 1510 |
| (2) | 3020 | 3020 |
| (3) | 21,750 | 5362 |
| | | 9892 |

Total of errors discovered *and corrected* is

$$9962 - 9892 = £70$$

The following items are found to be subject to compliance errors.

| | £ |
|---|---|
| (1) | 10,276 |
| (2) | 54,153 |
| (3) | 895 |
| (4) | 5364 |
| (5) | 412,000 |

6. Adjust for the monetary error.
   6.1. Calculate the estimated error values.

| | Overstatement |
|---|---|
| (1) | 2135 × 32,000 ÷ 30,326 = 2253 |
| (2) | 6827 × 32,000 ÷ 17,524 = 12,466 |
| (3) | 1000 × 32,000 ÷ 2756 = 11,611 |
| | **Understatement** |
| (1) | 1510 —[a] = 1510 |
| (2) | 3020 × 32,000 ÷ 3020 = 32,000 |
| (3) | 5362 × 32,000 ÷ 21,750 = 7889 |

[a] No estimate required here, the item containing the error exceeds the $J$ factor in value.

   6.2. Adjust the estimated error values for the precision adjustment factor from Table 17.1 on p. 204.

| | Estimated values ranked in value order (Overstatement) £ | Precision adjustment factor (86%) (Table 17.1) | £ |
|---|---|---|---|
| (1) | 12,466 | 1·51 = | 18,824 |
| (2) | 11,611 | 1·38 = | 16,023 |
| (3) | 2253 | 1·31 = | 2951 |
| | | | +37,798 |
| | (Understatement) | | |
| | | B/f | 37,798 |
| (1) | 32,000 | 0·14 = | 4480 |
| (2) | 7889 | 0·49 = | 3866 |
| (3) | 1510 | —[a] = | 1510 |
| | | | −9856 |
| | Total adjustment to MP | | +27,942 |

[a] This is the *actual* error in a block in excess of £32,000. It therefore needs no further adjustment. We cannot have underestimated it.

7. The figure for monetary precision must now be adjusted in accordance with the information derived from the sample.

| | £ |
|---|---|
| Initial MP | 64,000 |
| Add: adjustment for monetary error as calculated above | 27,942 |
| | 91,942 |
| Less: correction of monetary errors | 70 |
| | 91,872 |

Since the upper limit estimate on monetary precision exceeds £64,000 an extended audit is required, but the population is unlikely to be rejected without further test. However, some compliance errors were discovered and it may be that the initial estimate of an average internal control system is not supported by the evidence from the sample. The auditor's final step is, therefore, to examine the compliance errors to test the accuracy of the initial reliability factor used.

8. Analysis of compliance errors.

8.1. The value of the items in error are (£) 10,276 54,153 895 5364 412,000. Since the entire item is in error in each case the estimated value is equal to the *J* factor unless it exceeds the *J* factor when the actual amount is added in.

| Estimated value of item ranked in value order | PAF at 95% confidence level (Table 17.1) | Adjusted value |
|---|---|---|
| 412,000 | 1·00 | 412,000 |
| 54,153 | 1·00 | 54,153 |
| 32,000 | 1·75 | 56,000 |
| 32,000 | 1·56 | 49,920 |
| 32,000 | 1·46 | 46,720 |
| | | £618,793 |

8.2. Add three times the *J* factor to the figure calculated in 8.1

$$3 \times 32{,}000 + 618{,}793 = £714{,}793$$

8.3. Calculate the $G/V$ ratio

$$714{,}793/10{,}000{,}000 = 0{\cdot}072$$

8.4. From Figure 17.2 on p. 212 a $G/V$ ratio of 0·072 suggests a reliability factor of 3,[1] since an initial reliability factor of 2 was used this suggests that the initial assessment of the effectiveness of the internal control system was unduly optimistic. If we are to retain our original confidence level of 90%, i.e. $R$ of 2·0, a new precision limit, i.e. monetary precision, must be calculated. This is effected by using the formula

$$JR' = P'$$

where $J$ = $J$ factor, $R'$ = new reliability factor, $P'$ = new monetary precision.

Substituting the figures in the formula,

$$32{,}000 \times 3 = £96{,}000$$

## FINAL ESTIMATE OF ADJUSTED MP

We can now put together the various adjustments to the MP as follows:

| | |
|---|---|
| Original MP | 64,000 |
| Adjustment for | |
| compliance error | +32,000 |
| monetary error | +27,942 |
| correction of monetary errors discovered | −70 |
| Adjusted MP | £123,872 |

The auditor can now be 90% confident that the total value of error in the population is less than £123,872. But since his initial estimate of material error was £64,000 he is unlikely to accept this audit without a considerably more detailed check.

## THE ADVANTAGES OF MONETARY UNIT SAMPLING

The MUS system enjoys certain advantages and suffers from some disadvantages compared to the classical forms of sampling discussed earlier in this book. We will now examine some of these advantages and limitations.

[1] Our initial assessment of $R$ was based on *average* internal control system, i.e. section B.

## OPERATIONAL ADVANTAGES

The first advantage of the MUS system over the systems discussed previously is that less need be known about the population being audited. The number of units in the population and the standard deviation need neither be known nor estimated. The total value of the population needs to be known, but the value is almost invariably given in conventional audits. Also the population must be summed, but again this is conventional audit practice. No additional work is involved. The MUS system thus fits neatly into traditional auditing methodology.

The second advantage of the MUS system is that it forces the auditor to state the minimum unacceptable error value for the entire population. This is called monetary precision. Auditors have traditionally avoided the difficult problem of placing a specific numerical value on 'significant error' in an audit. The MUS system focusses attention on this important variable. This is, perhaps, the most important contribution of the MUS system towards improved audit practice.

A third advantage is the simplicity of the method of calculating sample size. No voluminous tables or complicated formulae are needed to calculate sample size. The sample size is calculated by dividing the reliability factor (Poisson factor) by the percentage of total value which represents the minimum significant error.

The method of estimating the efficiency of internal control in the initial stages of the audit and then testing the accuracy of this estimate after the sample has been drawn has much to recommend it. A scientific method of evaluating internal control has long eluded the auditor.

Finally a scientific audit method that, in effect, uses the trial balance as the document to be audited has the not inconsiderable advantage of providing the auditor with a complete overall view of the audit at all times. The method overtly demonstrates that the audit effort is being allocated in proportion to the value of the various subpopulations making up the trial balance.

## THEORETICAL ADVANTAGES

The previous section discussed certain operational advantages which the MUS system enjoys over other statistical sampling methods. In addition to these operational advantages the MUS system enjoys certain theoretical advantages, in the sense that the theory behind the system appears to be superior to some of the rival systems discussed previously.

The most obvious theoretical advantage lies in the way the MUS system handles the extreme skewness inherent in almost all accounting populations. A physical unit sampling system must face up to the formidable problem of estimating the standard deviation of the error value in each stratum of the stratified population. The MUS system handles the problem by, in effect, treating each $J$ block as a binominal population having a fraction of 'tainted' pounds. The calculation of an upper bound on this tainted percentage presents a much simpler problem.

A second theoretical advantage lies in the fact that the sample size is based on the total value of the audit. That is the sum of both sides of the trial balance. Thus the number of units drawn from any component subpopulation, such as sales or inventory, is likely to be quite small. This is theoretically acceptable since the sample frame is the total audit value. Other systems of statistical sampling treat each item in the trial balance as a separate population thus generating a very large total sample. This latter sample is likely to provide an overall level of confidence which is higher than the *minimum* level required by the auditor, although it may be that these same systems may over-sample units of small value to such an extent that the underestimate of the level of confidence may not, in fact, be so substantial.

The Bayesian approach of using a prior probability distribution based on the estimated efficiency of the internal control system to decide on the required level of confidence in the inference from the sample is a useful contribution to audit theory. The method applies a numerical framework to a judgemental factor and so permits comparisons between different audits' (and auditors') evaluation of internal control. The additional benefit of testing this estimate later in the audit has already been discussed.

Finally the derivation of sample size from a comparison of total audit value to minimum unacceptable error value provides a proper strategic framework for the audit. The key factor in auditing is value not number of units audited. The MUS system places value at the heart of the audit.

## LIMITATIONS OF THE MUS SYSTEM

The MUS is an admirable system which, as we have seen, provides significant advantages to the external auditor compared to alternative systems. The method, however, suffers from certain limitations and, as its advocates admit, can still be improved.

The first limitation arises from the fact that the system is handmade for the *external* auditor. The method only works when a grand total is known and the error value must be limited to below a given figure.

Earlier in this book we noted examples of applications of statistical sampling to audit and accounting work where these conditions would not hold. It may be that the total of an accounting population or the error value are the very facts we wish to estimate. Again in internal auditing it is often the number rather than the value of errors that we need to estimate for control purposes. Or, alternatively, we may wish to control the error values in each subpopulation, creditors, vehicles, etc., independently of the other subpopulations, that is we are not looking at the audit as a whole in deciding on our MP.

In summary, the MUS system was specifically designed to assist the external auditor in performing a total audit. The method is of limited value in other uses.

The methods of scientific sampling described elsewhere in this book are based on classical statistical theory. The statistical theory behind the MUS system is more complex and therefore, more difficult to explain to the non-statistician. This is a distinct disadvantage since an auditor who is not clear on why he can draw a given conclusion may, quite inadvertently, draw an invalid conclusion. The MUS system mixes objective with subjective evaluations and the manuals explaining the system do not lay sufficient stress on those points where the subjective intuitive element is present. It is curious to note that no mathematical exposition of the MUS system appeared between 1961, the year of its inception, and 1973. This does not suggest that we doubt its theoretical validity—this has been checked out by several eminent statisticians—but we think it a pity that more emphasis was not laid on exposing the theoretical foundations of the system, which are unfamiliar, but not difficult to grasp.

The MUS system described in this chapter is based on controlling the *overstatement* of an amount. In Table 17.1 on p. 204 the overstatements are increased in value to ensure they are not understated, the understatements, in the lower half of the table, are *reduced* in value to ensure that they are not overstated. This is only a conservative policy if the auditor considers that minimizing understatement is conservative. One school of thought considers that both understatements and overstatements are equally important and that both should be multiplied by the factor in the *top* half of Table 17.1. This will, of course, tend to reduce the error value if error is a positive amount, which it usually is.

A not dissimilar problem arises from the assumption, built into the MUS system, that the maximum value of an overstatement is the value of the item. This is only true if credit balances do not exist in what are normally debit balances, and vice versa. Our research suggests that this is not a serious problem since *large* credit balances rarely occur among debit balances and vice versa. However, the possibility exists and should be noted.

A more serious problem is the zero balance. The MUS system cannot select a zero balance, yet the zero balance could be an error concealing a large positive balance, i.e. an understatement. We suggest that if a large number of zero balances exist, a sample of these should be checked independently of the MUS system.

Another limitation of the MUS system is that the probability of picking up a 'tainted' pound is affected by the distribution of the 'tainted' pounds through the population. In the extreme case an error value of £10,000 could be distributed as one unit of £10,000 or 10,000 units of one pound each. Clearly the probability of picking up an account containing a tainted pound is very much greater in the latter case. This fact, by itself, would not be important if it were not that the MUS system has a built in conservative bias, similar to that explained in the chapter on discovery sampling.

The reader will recall that the trouble with discovery sampling is that the system not only suggests rejection at the minimum unacceptable error rate (MUER) but also frequently suggests rejection at error rates well *below* the MUER. Discovery sample sizes are calculated without taking into account the maximum unacceptable rejection rate (MURR). Acceptance sampling takes both the MUER and the MURR into account.

The MUS system is likely to work satisfactorily so long as the error value is distributed among only a few items. If it is distributed among a large number of items the probability of picking up one 'tainted' pound is very high and so the total value of error in the population is likely to be overestimated since if even one 'tainted' pound is discovered the estimate of overstatement error is assumed to be at least equal to the MP.

For example if a population of £10,000,000 has an MP of £10,000 and the actual error value is £8,000 the population ought to be accepted. If we assume a confidence level of 80% the sample size is 200. The £8000 could be a single account in error, that is 8000 100% tainted pounds or, say, 100 accounts each containing £1600 of error value all 5% tainted (5% × 1600 × 100 = £8000). In the latter case £160,000 of the £1,000,000 will throw out a tainted pound, in the former case only £8000 of the £1,000,000 will do so. Thus if the error value is widely distributed in lightly tainted pounds the MUS system is likely to grossly overestimate the actual error rate in the population.

The MUS system is thus too conservative. It will tend to reject many populations which ought to be accepted. This is not a serious criticism only if the MP is well above the actual error value in the population.

The effect of this conservative bias in the MUS system is that many audit populations which are, in fact, acceptable may be rejected and additional, unnecessary audit work carried out upon them. Alternatively the auditor, faced with many false alarms, may become disillusioned with the system.

The advocates of the MUS system claim that this theoretical criticism has not proved a drawback in practice since value errors are rarely discovered.

The evaluation of the MP, the minimum unacceptable error value, is another unsatisfactory feature of the MUS system. Various users of the MUS system supplied the following explanations to the author as to how they estimated the MP.

1. 'The minimum error significant to a financial analyst examining the accounts.'
2. 'The minimum amount of error or defalcation which would cause concern to the top management of the company.'
3. 'The maximum amount of error value acceptable, taking into account the rather arbitrary valuation methods used in valuing inventory and other assets.'
4. '1% of sales.'
5. '2% of net asset value.'
6. 'Based on totals of the trial balance—say 0·1% of this.'
7. 'Depends on type of business, how accurate the accounts are.'

The figure of 1% of sales turnover for the period was the most popular rule of thumb method suggested. We feel that a more scientific method of deciding on the MP ought to be worked out. Finally we would fault the MUS system for being rather vague on the action to take if the population is rejected.[1] In theory, if the MP is the minimum unacceptable error value then the discovery of *any* monetary error should result in the population being rejected. In practice if the MP is only exceeded by a small margin the population is accepted. We suggest that it might be better to decide initially on two upper limits, an expected error value, which becomes the MP and a higher minimum unacceptable error value (MUEV). Only if the MUEV is exceeded should the population be subjected to further

[1] K. Stringer has suggested that the auditor should set up a reserve equal to the excess over the MP.

audit work. Certainly the action required when the MP is exceeded should be handled in a more scientific manner.

The criticism of the MUS system set out above should not be interpreted as a rejection of the system. The MUS system is, in our opinion, the best system so far devised for external auditing; it is specifically designed to meet the needs of the professional auditor. It is, however, still open to improvement, and our comments are made in a spirit of constructive criticism.

We believe that the form of statistical auditing eventually adopted by the profession will prove to be a modified version of the MUS system.

## QUESTION SERIES 17

1. Calculate the sampling interval $J$ and the sample size in the following examples.

| | $V$ | $P$ | Confidence level |
|---|---|---|---|
| (a) | 500,000 | 10,000 | 86% |
| (b) | 50,000,000 | 50,000 | 63% |
| (c) | 1,500,000 | 10,000 | 90% |

(*)

2. With a $J$ factor of £42,000 the following monetary errors are found.

| Item value (£) | Error value (£) |
|---|---|
| Overstatement | |
| 32,365 | 12,174 |
| 18,326 | 18,326 |
| 50,000 | 10,000 |
| Understatement | |
| 8354 | 927 |
| 39,465 | 3156 |

Calculate the adjustment required to the monetary precision. Population value £5 million. Confidence level 63% MP £42,000. (*)

3. Facts as for problem 2. In addition the following compliance errors are found.

Value of items containing compliance error

| £ |
|---:|
| 53,825 |
| 2700 |
| 7600 |
| 23,576 |

Does this value of compliance error suggest that the original reliability factor based on a good internal control system, needs to be revised? (*)

## SOME ANSWERS TO QUESTION SERIES 17

1. (a) £5000 100
   (b) £50,000 1000
   (c) £4348 344
2. Approximately £72,000 should be added to the MP of £42,000.
3. Yes. The reliability factor was 1·0 (63%) the evidence from the sample suggests it ought to be 1·1. (The $G/V$ ratio is 0·076, giving, via Figure 7.2, a new $R$ factor of 1·1.)

# 18

# Some Statistical Sampling Tables Available

## INTRODUCTION

By using the formula set out in Chapter 19 an auditor can calculate the required sample size under given conditions. In some cases it is relatively easy to calculate sample size from the formula. Examples of easy calculation are estimation sampling of values (variables) and calculating the standard error on a proportion.

In other cases statisticians have provided tables and graphs to assist the rapid calculation of sample size. Examples are provided in Tables 21.1 to 21.4 on pp. 238–260.

However in other cases the volume of calculation generated by the formula is such that it will test the largest computer.

If a professional firm wishes to delegate statistical sampling to junior staff it is best to provide a comprehensive set of precalculated tables for finding sample size. Many such tables are available. In this chapter I will describe some of the more accessible sets of tables.

## DESCRIPTION OF SAMPLING TABLES AVAILABLE

Brown and Vance (1961) is a useful set of tables for estimating error rates and other proportions, that is it is useful in estimation sampling of attribute problems.

The tables provide the following:

| | |
|---|---|
| Confidence levels | 90%, 95%, 99%, 99·7% |
| Precision limits ± | 0·5% to 10% |
| Population size | 200 to 1,000,000 |

Arkin (1963), Table F, provides a useful set of tables for estimating population proportions from sample proportions.

| | |
|---|---|
| Population size | 500 to 100,000 |
| Percentage in sample | 0% to 50% |
| Sample size | 50 to 3000 |
| Confidence levels | 90%, 95%, and 99% |

Arkin (1963), Table E, also provides a very comprehensive set of tables for calculating sample size in estimation sampling of values (variables).

| | |
|---|---|
| Population size | 500–1,000,000 |
| Confidence level | 90%, 95%, 99%, 99·9% |

A very large number of tables have been compiled for acceptance sampling purposes. Again Arkin (1963), Table K, provides acceptance sampling plans in a format which makes it easy to test the MUER and MURR simultaneously.

| | |
|---|---|
| Batch size | 200 to 50,000 |
| Error rate | 0·05% to 15% |
| Acceptance number | 0 to 5 |
| Sample size | up to 800 |

The best known acceptance sampling tables are those compiled by Dodge and Romig (1944). These use confidence levels of 90% on both MUER and MURR. The United States Defence Department (1950) have also compiled a very comprehensive set of acceptance sampling tables. A useful set of stop–go tables has been compiled by the U.S. Air Force (1960). The Institute of Internal Auditors (U.S.A.) (1970) have also compiled a set of stop–go tables for auditing purposes as well as many other useful tables.

The most useful set of discovery sampling tables for auditors is to be found in Arkin (1963), Table J.

| | |
|---|---|
| Population size | 200–200,000 |
| MUER | 0·01% to 15% |

Kraft (1968) provides a set of tables for calculating the joint probability when using Bayesian statistics in auditing.

| | |
|---|---|
| Sample size | 100, 150, 200. |
| Errors discovered in samples | 0 to 5. |

An optimistic, average, and pessimistic prior probability is catered for.

# 19

# A Summary of the Formulae

## INTRODUCTION

In this chapter we will summarize the formula presented elsewhere in this book.

1. The standard error of the sample estimate of a proportion is given by the formula;

$$s = \left(\frac{p(1-p)}{n}\right)^{1/2}\left(1-\frac{n}{N}\right)^{1/2}$$

where $s$ = standard error of a proportion, $p$ = percentage of sample with condition, $n$ = sample size, $N$ = population size.

This formula becomes increasingly inaccurate as $p$ approaches 0%. Below about 3% it is advisable to use corrected tables such as Arkin (1963), Table F.

2. The formula to determine sample size in estimation sampling of attributes is:

$$n = \frac{p(1-p)}{(e/f)^2 + p(1-p)/N}$$

where $n$ = required sample size, $p$ = expected percentage of sample with condition, $e$ = required precision limit, $\pm e\%$, $f$ = factor decided by confidence level (90% = 1·64, 95% = 1·96, 99% = 2·58, 99·7% = 2·97), $N$ = population size.

Remember to make a conservative estimate of $p$.

3. The formula to determine sample size in estimation sampling of variables (values) is:

$$n = \frac{S^2}{s^2 + S^2/N}$$

where $n$ = sample size, $S$ = standard deviation (estimated) of population, $s$ = standard error of sample means, $N$ = population size.

Note that $s$ is usually calculated by means of making the required unit precision limit, $p$, equal to confidence level factor, $c$, of

$$\begin{aligned} 90\% &= 1{\cdot}64 \\ 95\% &= 1{\cdot}96 \\ 99\% &= 2{\cdot}58 \\ 99{\cdot}7\% &= 2{\cdot}97 \end{aligned}$$

times $s$. I.e.

$$cs = p.$$

$$s = \frac{p}{c}.$$

4. To calculate the discovery and acceptance sampling tables we need to use some variant of the hypergeometric distribution formula.

This is,

$$r = \left(\frac{a!}{b!(a-b)!}\right)\left(\frac{(N-a)!}{(n-b)!\{(N-a)-(n-b\}!}\right)\Big/\left(\frac{N!}{n!(N-n)!}\right)$$

$r$ = probability of this situation occurring, $a$ = number of units in population with given condition, $b$ = number of units in sample with given condition, $n$ = number of units in the sample, $N$ = number of units in the population.

When the population is infinite, i.e. we replace all samples drawn, we can use the well known binomial formula.

$$r = \frac{n!}{b!(n-b)!}p^b(1-p)^{n-b}$$

where $r$ = probability of this situation occurring, $n$ = number of units in the sample, $b$ = number of units in sample with specified condition, $p$ = percentage of population with specified condition, ! = factorial, i.e. $3! = 3 \times 2 \times 1 = 6$.

In all auditing applications the sample size is so large that the required computations are enormous. Therefore the auditor must use precomputed tables such as Arkin (1963), Tables J and K.

5. Standard deviation. The crudest formula for calculating the standard deviation is:

$$s = \left\{\frac{\sum_{i=1}^{h}(v_i - \bar{x})^2 f_i}{n}\right\}^{1/2}$$

where $s$ = standard deviation, $n$ = number of units in population, $v$ =

value of reading, $\bar{x}$ = mean of population, $f$ = number of units in given class, $\sum$ = sum of.

If a desk calculator or computer is available a more suitable formula is:

$$s = \left\{\frac{\sum (v)^2}{n} - \left(\frac{\sum v}{n}\right)^2\right\}^{1/2}$$

where $s$ = standard deviation, $v$ = value of reading, $n$ = number of units in the population, $\sum$ = sum of.

The formula for estimating the standard deviation using the average range method is:

$$s = \frac{\sum r}{n.c}$$

where $s$ = standard deviation, $\sum r$ = total of group ranges, $n$ = number of groups, $c$ = coefficient determined from Table 21.6 on p. 265.

6. Stratification. The formula for calculating the stratification coefficient is:

$$t = \frac{\sum (ns)}{N^2(cp)^2 + \sum (ns^2)}$$

where $t$ = stratification coefficient, $n$ = number of units in stratum, $s$ = standard deviation of each stratum, $p$ = required unit precision limit, $N$ = population size, $c$ = 0·606 for 90% confidence level, 0·510 for 95% confidence level, 0·388 for 99% confidence level.

Having calculated $t$ we use the following formula to calculate the size of sample to draw from each stratum to meet the required conditions.

$$n \times s \times t = r$$

where $n$, $s$ and $t$ are as above and $r$ is the required sample size.

7. Multistage sampling. To calculate the sampling error of the mean estimate from a multistage sample we must add together the sampling error at each stage. For a two stage sample the standard error is given by the formula:

$$s = \left(r^2 + \frac{n}{N}a\right)^{1/2}$$

where $s$ = sampling error of mean estimate of total, $r$ = sampling error of the secondary sample averages for the primary sample averages, $a$ = average of the squares of the individual sampling errors of the primary sampling units, $n$ = number of primary sampling units sampled, $N$ = number of primary sampling units.

8. Finite population correction factor. Once the calculation is completed for a population of infinite size the result should be multiplied by the following formula to adjust it to the actual size of the population.

$$f = \left\{1 - \frac{n}{N}\right\}^{1/2}$$

where $f$ = finite population correction factor, $n$ = sample size, $N$ = population size.

9. Mean, median and mode. The mean is given by the formula

$$m = \sum_{i=1}^{h} \frac{v_i f_i}{n}$$

where $m$ = mean, $v$ = value of reading, $f$ = number of units of this value, $n$ = number of units in population.

The median is the $\frac{1}{2}(n + 1)$th unit in the population when the units are ranked ordinally by value.

The mode is the most frequent value found in the population.

10. Skewness. There are two formulae available for measuring the skewness of a population:

$$k = \frac{(m - d)}{s} \quad \text{and} \quad k = 3\,\frac{(m - n)}{s}$$

where $k$ = skewness, $m$ = mean, $d$ = mode, $s$ = standard deviation, $n$ = median.

If the population is classified it is easy to estimate the median. If not it is easier to estimate the mode.

11. Measuring the randomness of a series of numbers. Having estimated the median and underlined all items above the median, the data for measuring randomness is as follows

$$m = (n + 1)$$

$$s = \left\{\frac{2n(n - 1)}{2n - 1}\right\}^{1/2}$$

where $n$ is the number of units underlined.

The confidence limits on the mean are calculated as follows:

$$f = c.s$$

where $f$ = confidence limit, $s$ = standard deviation, $c$ = 1·64 for 90% level of confidence, 1·96 for 95% level of confidence, 2·58 for 99% level of confidence.

# 20

# A Selective Bibliography on Statistical Sampling in Auditing


### BOOKS ON THE GENERAL THEORY OF SAMPLING

Slonim, M. J. *Guide to Sampling*, Pan Piper, London (1968). (An elementary introduction to the basic principles of sampling.)

Hoel, P. G. *Elementary Statistics*, John Wiley, New York (1960).

Stuart, A. *Basic ideas of Scientific Sampling*, Griffin, London (1962). (Never was so much information about the basic principles of sampling packed into so short a space. Non-mathematical but quite tough reading.)

Cochran, W. G. *Sampling Techniques*, John Wiley, New York (1963). (This is, perhaps, the classic treatment of the subject. Quite difficult for an accountant not well versed in mathematics.)

Yates, F. *Sampling Methods for Censuses and Surveys*, Griffin, London (1960). (A most comprehensive introduction to the practical problems of designing sample surveys.)

### BOOKS ON APPLYING SAMPLING TO AUDITING AND ACCOUNTING

Hill, Henry P., Joseph, L. Roth, and Arkin, Herbert. *Sampling in Auditing: A simplified Guide and Statistical Tables*, Ronald Press Company, New York (1962). (Probably the best basic introduction to what it is all about. But the book does not attempt to delve very deeply.)

Trueblood, R. M., and Cyert, R. M. *Sampling Techniques in Accounting*, Prentice Hall, Englewood Cliffs, New Jersey (1957). (An early book, but clearly written with some good examples.)

Cyert, R. M., and Davidson, H. Justin. *Statistical Sampling for Accounting Information*, Prentice Hall, Engelwood Cliffs, New Jersey (1962). (Strong emphasis on theory of sampling.)

Vance, Lawrence L., and Neter, John. *Statistical Sampling for Auditors and Accountants*, John Wiley, New York (1956). (Strong on acceptance sampling and includes useful list of early applications of the technique.)

Arkin, Herbert. *Handbook of Sampling for Auditing and Accounting*, Vol. 1, Methods, McGraw–Hill, New York (1963). (A lucid non-mathematical introduction to sampling theory with emphasis on auditing applications. Written by a well known applied statistician. A most useful set of sampling tables are provided. These tables have proved a major contribution to popularizing the use of statistical sampling methods in auditing.)

Brown, R. Gene, and Vance, Lawrence L. *Sampling Tables for Estimating Error Rates or Other Proportions*, Institute of Business and Economic Research, University of California, Berkeley, California (1961). (A comprehensive set of tables for estimating proportions. Very clearly set out.)

## ARTICLES

I estimate that at the date these words are written (1974) something of the order of four hundred articles have been published on the topic of applying statistical sampling to audit and control procedures. I do not claim to have read all of these.

The following list includes some of the more interesting, original and useful articles, which I have read.

### 1. *The theory of sampling as applied to auditing*

1.1 'Accuracy in statistical sampling'. H. J. Davidson. *Accounting Review*, **34** (1959), 356–365.

1.2 'Random samples in audit tests'. R. W. Johnson. *Journal of Accountancy*, **104** (December 1957), 43.

1.3 'Statistical sampling tables for auditors'. R. Gene Brown. *Journal of Accountancy*, **111** (May 1961), 46–54.

1.4 'On a mixed sequential estimating procedure with application to audit tests in accounting'. A. Charnes, H. J. Davidson and K. Kortanek. *Accounting Review*, **39** (1964), 241–250.

1.5 'Some observations on statistical sampling in auditing'. H. F. Stettler. *Journal of Accountancy*, **121** (April 1966), 55–60.

### 2. *Estimation sampling of attributes*

2.1 'Auditing voluminous data by modern sampling methods'. D. D. Davis and A. Rounsaville. *Journal of Accountancy*, **107** (June 1959), 45–51.

2.2 'An application of statistical techniques in internal auditing'. J. B. Craig. *Journal of Accountancy*, **108** (November 1959), 39–45.

### 3. *Estimation sampling of variables*

3.1 'Inductive Accounting—an application of statistical sampling techniques'. W. C. Dalleck. *O.R. Applied Special Report*, No. 17, American Management Association, New York.
(Examines the method by which inter-airline debt is allocated on a sampling basis.)

3.2 'Inventory determination by means of statistical sampling when clients have perpetual records'. W. D. Hall. *Journal of Accountancy*, **123** (March 1967), 65–71.

3.3 'A case study of statistical sampling'. R. F. Obrock. *Journal of Accountancy*, **105** (March 1958), 53–59.
(Discusses stratification.)

3.4 'Statistical sampling technique in the aging of accounts receivable in a department store'. R. M. Cyert and R. M. Trueblood. *Management Science*, **3**, 2 (1957), 185–195.

3.5 'Estimating the liability for unredeemed stamps'. H. J. Davidson. *Journal of Accounting Research*, **5**, 2 (1967), 186–207.

3.6 'Scientific test of sampling applied to audit test of inventories'. R. Taussig. *N.A.A. Bulletin* (January 1960), 21–28.

3.7 'Applicability of statistical sampling techniques to the confirmation of accounts receivable'. J. Neter. *Accounting Review*, **31** (January 1956), 82–94.

3.8 'Can scientific sampling techniques be used in railroad accounting?' C. W. Churchman. *Railway Age* (June 9, 1952), 61–64.
(Discusses interline charges.)

3.9 'Lifo and statistical sampling'. H. J. Davidson and R. J. Monteverde. *Management Science* (April 1959), 279–292.

3.10 'Applied sampling doubles inventory accuracy, halves cost'. A. Rudell. *N.A.A. Bulletin*, **39**, 2 (October 1957), 5–11.

4. *Acceptance sampling*

4.1 'An auditors approach to sampling'. K. A. Sherwood. *The Accountant* (May 2, 1964).

4.2 'Sampling in auditing—a case study'. H. G. Trentin. *Journal of Accountancy*, **125** (March 1968), 39–43.

4.3 'Some notes of reservation on the use of sampling tables in auditing'. R. J. Monteverde. *Accounting Review*, **30** (October 1955), 582–591.

5. *Discovery sampling*

5.1 'Discovery sampling in auditing'. H. Arkin. *Journal of Accountancy*, **111** (February 1961), 51–54.

6. *Practical advice in applying statistical sampling to auditing*

6.1 'Statistical sampling techniques as an audit tool'. J. Smurthwaite. *Accountancy*, **76**, 859 (March 1965), 201–209.
(An account of one auditor's experiences in using statistical sampling.)

6.2 'Statistical sampling: a practical approach'. T. G. Haworth. *Accountancy*, **80** (February 1969), 101–109.
(Provides useful graphs for selecting sample size in estimating error rates. His criticism of Arkin's tables is not correct.)

6.3 'Some practical problems in applying statistical sampling in auditing'. T. W. McRae. *Accountants Magazine*, **75**, 781 (July 1971), 369–377.

6.4 'An experimental study of audit confirmations'. G. B. Davis and R. R. Palmer. *Journal of Accountancy* (June 1967), 36–40.

6.5 'Statistical v. judgement sampling'. H. F. Aly and J. I. Duboff. *Accounting Review*, **46**, 1 (1971), 119–128.
(Empirical study of auditing debtors in a small retail store.)

6.6 'Review of developments in statistical sampling for accountants'. Lawrence L. Vance. *Accounting Review*, **35** (January 1960), 19–28.
(An interesting review of the literature up to 1960.)

7. *Audit sampling by computer*

7.1 'Case study on the use of computer and statistical techniques'. J. B. Irvine. *Journal of Accountancy*, **117** (April 1964), 67–68.

8. *The use of Bayesian Statistics in auditing.*

8.1 'The new business statistics'. H. V. Roberts. *Journal of Business*, (January 1960), 29.
(A good introduction. No auditing mentioned.)

8.2. 'Statistical sampling for auditors. A new look'. W. H. Kraft. *Journal of Accountancy*, **126** (August 1968), 49–56.
(Important primal article on the use of BS in auditing. Useful tables.)

8.3 'Bayesian statistical methods in auditing'. J. A. Tracy. *Accounting Review*, **44**, 1 (1969), 90–98.
(A clear explanation of difference between classical and Bayesian approach.)

8.4 'Bayesian analysis in auditing'. J. E. Sorensen. *Accounting Review*, **44**, 3 (1969), 555–561.
(Attempts to measure economic viability of increasing size of audit sample.)

9. *Sampling and the law*

9.1 'The admissability of sample data to a court of law'. R. C. Sprowels. *University of California Law Review* (February 1957), 224.

10. *Uses of statistical sampling by government agencies*

10.1 'Use of statistical techniques by the government'. L. L. Teitelbaum and N. L. Burton. *Journal of Accountancy*, **110** (November 1960), 24–29.

11. *The case against statistical sampling*

11.1 'An accountant looks at statistics'. H. P. Hill. *Journal of Accountancy*, **105** (April 1958), 57–65.

11.2 'Statistical sampling and the auditor' (section only). T. W. McRae. *The Accountants Magazine* **69**, 707 (May 1965), 390–395.

11.3 Letter from W. H. Whitney. *Journal of Accountancy*, **106** (1958), 23.

12. *Official pronouncements on statistical sampling in auditing*

12.1 'Statistical sampling and the independent auditor'. Report by AICPA. *Journal of Accountancy*, **114** (February 1962), 60–62.

12.2 'Relationship of statistical sampling to generally accepted auditing standards'. Report by AICPA. *Journal of Accountancy* (July 1964), 56.

# 21

# Tables and Graphs for Calculating Sample Size Plus Set of Instructions for Operating Each System

SAMPLING TABLE INDEX

## AREAS UNDER THE NORMAL CURVE

Table 21.1 allows the sampler to transform a given standard deviation ($z$) into a probability. The table gives values between $z = 0$ and positive values of $z$. To find two tailed values double this figure. To find one tailed values add 50%.

*Example.* What probability is represented by 1·96 standard deviations either side of the mean?

From the table we find that 1·96 in the $z$ column gives a figure of 0·4750. $2 \times 0{\cdot}4750 = 0{\cdot}95$, which is a probability of 95%.

Table 21.1. Areas of a standard normal distribution. An entry in the table is the proportion under the entire curve which is between $z = 0$ and a positive value of $z$.

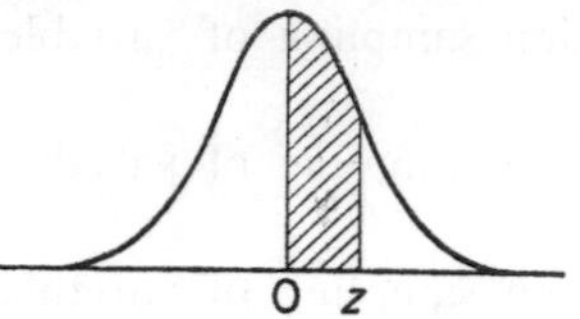


| $z$ | ·00 | ·01 | ·02 | ·03 | ·04 | ·05 | ·06 | ·07 | ·08 | ·09 |
|---|---|---|---|---|---|---|---|---|---|---|
| 0·0 | ·0000 | ·0040 | ·0080 | ·0120 | ·0160 | ·0199 | ·0239 | ·0279 | ·0319 | ·0359 |
| 0·1 | ·0398 | ·0438 | ·0478 | ·0157 | ·0557 | ·0596 | ·0636 | ·0675 | ·0714 | ·0753 |
| 0·2 | ·0793 | ·0832 | ·0871 | ·0910 | ·0948 | ·0987 | ·1026 | ·1064 | ·1103 | ·1141 |
| 0·3 | ·1179 | ·1217 | ·1255 | ·1293 | ·1331 | ·1368 | ·1406 | ·1443 | ·1480 | ·1517 |
| 0·4 | ·1554 | ·1591 | ·1628 | ·1664 | ·1700 | ·1736 | ·1772 | ·1808 | ·1844 | ·1879 |
| 0·5 | ·1915 | ·1950 | ·1985 | ·2019 | ·2054 | ·2088 | ·2123 | ·2157 | ·2190 | ·2224 |
| 0·6 | ·2257 | ·2291 | ·2324 | ·2357 | ·2389 | ·2422 | ·2454 | ·2486 | ·2517 | ·2549 |
| 0·7 | ·2580 | ·2611 | ·2642 | ·2673 | ·2703 | ·2734 | ·2764 | ·2794 | ·2823 | ·2852 |
| 0·8 | ·2881 | ·2910 | ·2939 | ·2967 | ·2995 | ·3023 | ·3051 | ·3078 | ·3106 | ·3133 |
| 0·9 | ·3159 | ·3186 | ·3212 | ·3238 | ·3264 | ·3289 | ·3315 | ·3340 | ·3365 | ·3389 |
| 1·0 | ·3413 | ·3438 | ·3461 | ·3485 | ·3508 | ·3531 | ·3554 | ·3577 | ·3599 | ·3621 |
| 1·1 | ·3643 | ·3665 | ·3686 | ·3708 | ·3729 | ·3749 | ·3770 | ·3790 | ·3810 | ·3830 |
| 1·2 | ·3849 | ·3869 | ·3888 | ·3907 | ·3925 | ·3944 | ·3962 | ·3980 | ·3997 | ·4015 |
| 1·3 | ·4032 | ·4049 | ·4066 | ·4082 | ·4099 | ·4115 | ·4131 | ·4147 | ·4162 | ·4177 |
| 1·4 | ·4192 | ·4207 | ·4222 | ·4236 | ·4251 | ·4265 | ·4279 | ·4292 | ·4306 | ·4319 |
| 1·5 | ·4332 | ·4345 | ·4357 | ·4370 | ·4382 | ·4394 | ·4406 | ·4418 | ·4429 | ·4441 |
| 1·6 | ·4452 | ·4463 | ·4474 | ·4484 | ·4495 | ·4505 | ·4515 | ·4525 | ·4535 | ·4545 |
| 1·7 | ·4554 | ·4564 | ·4573 | ·4582 | ·4591 | ·4599 | ·4608 | ·4616 | ·4625 | ·4633 |
| 1·8 | ·4641 | ·4649 | ·4656 | ·4664 | ·4671 | ·4678 | ·4686 | ·4693 | ·4699 | ·4706 |
| 1·9 | ·4713 | ·4719 | ·4726 | ·4732 | ·4738 | ·4744 | ·4750 | ·4756 | ·4761 | ·4767 |
| 2·0 | ·4772 | ·4778 | ·4783 | ·4788 | ·4793 | ·4798 | ·4803 | ·4808 | ·4812 | ·4817 |
| 2·1 | ·4821 | ·4826 | ·4830 | ·4834 | ·4838 | ·4842 | ·4846 | ·4850 | ·4854 | ·4857 |
| 2·2 | ·4861 | ·4864 | ·4868 | ·4871 | ·4875 | ·4878 | ·4881 | ·4884 | ·4887 | ·4890 |
| 2·3 | ·4893 | ·4896 | ·4898 | ·4901 | ·4904 | ·4906 | ·4909 | ·4911 | ·4913 | ·4916 |
| 2·4 | ·4918 | ·4920 | ·4922 | ·4925 | ·4927 | ·4929 | ·4931 | ·4932 | ·4934 | ·4936 |
| 2·5 | ·4938 | ·4940 | ·4941 | ·4943 | ·4945 | ·4946 | ·4948 | ·4949 | ·4951 | ·4952 |
| 2·6 | ·4953 | ·4955 | ·4956 | ·4957 | ·4959 | ·4960 | ·4961 | ·4962 | ·4963 | ·4964 |
| 2·7 | ·4965 | ·4966 | ·4967 | ·4968 | ·4969 | ·4970 | ·4971 | ·4972 | ·4973 | ·4974 |
| 2·8 | ·4974 | ·4975 | ·4976 | ·4977 | ·4977 | ·4978 | ·4979 | ·4979 | ·4980 | ·4981 |
| 2·9 | ·4981 | ·4982 | ·4982 | ·4983 | ·4984 | ·4984 | ·4985 | ·4985 | ·4986 | ·4986 |
| 3·0 | ·4987 | ·4987 | ·4987 | ·4988 | ·4988 | ·4989 | ·4989 | ·4989 | ·4990 | ·4990 |

## DISCOVERY SAMPLING TABLES

Using Table 21.2.

1. Decide on:

Minimum unacceptable error rate.
Level of confidence required in the inference.
Size of population.

2. Go to Table 21.2. Find given population block of numbers.
3. Within this block find the required MUER and confidence level.
4. At the juncture of this row and column is given the required sample size.
5. Draw a random sample of this size.
6. If no errors are found there is a $p\%$ probability that the error rate in the population is less than the percentage given in the MUER column.

Table 21.2. Discovery sampling table. If sample of size $n$ is drawn from the population and *no* errors are discovered the auditor can be $c\%$ confident that the error rate in the population is *less* than $x\%$. MUER = minimum unacceptable error rate. (The figures are derived by taking the most conservative of two approximation methods.)

| Population | MUER | Level of confidence | | | |
|---|---|---|---|---|---|
| | | 80% | 90% | 95% | 99% |
| 500 | 2 | 73 | 103 | 134 | 197 |
| | 1 | 130 | 185 | 240 | 353 |
| | 0·5 | 220 | 320 | 416 | 611 |
| 1000 | 2 | 74 | 107 | 140 | 204 |
| | 1 | 145 | 205 | 266 | 391 |
| | 0·5 | 275 | 370 | 481 | 707 |
| 2000 | 2 | 75 | 110 | 143 | 210 |
| | 1 | 152 | 220 | 286 | 420 |
| | 0·5 | 298 | 410 | 533 | 783 |
| 5000 | 2 | 77 | 112 | 146 | 214 |
| | 1 | 154 | 230 | 300 | 439 |
| | 0·5 | 308 | 440 | 572 | 840 |
| 10,000 | 2 | 79 | 114 | 148 | 220 |
| | 1 | 157 | 235 | 305 | 449 |
| | 0·5 | 316 | 460 | 598 | 878 |
| Infinite | 2 | 80 | 115 | 150 | 230 |
| | 1 | 160 | 240 | 310 | 460 |
| | 0·5 | 320 | 475 | 600 | 920 |

## ESTIMATION SAMPLING OF ATTRIBUTES

Using the graphs.

1. Decide on:

   Expected error rate (or other proportion).
   Maximum acceptable error rate.
   Level of confidence required in the inference from sample.

2. If error rate, etc. is 5% or less, use Figures 21.1.1 to 21.1.3.
   2.1 Go to table providing required level of confidence.
   2.2 Find expected error rate line on this table.
   2.3 Move up or down this expected error rate line until you are immediately above maximum acceptable error rate on horizontal axis.
   2.4 The vertical axis immediately across from this point will give you the required sample size. If population smaller than 10,000 multiply sample size by finite population correction factor from Table 21.7.
   2.5 Draw a random sample of this size from population.
   2.6 Calculate error rate in sample.
   2.7 Return to same table for second time. Find that maximum error rate in population that corresponds to given sample size and actual error rate in sample.
   2.8 If maximum error rate in population acceptable accept population if not reject.
3. If proportion is above 5%, use Figures 21.2.1 to 21.2.3.
   3.1 Decide on level of confidence required.
   3.2 Find table providing that level of confidence.
   3.3 Draw random sample of 100 and calculate proportion of this sample having condition sought.
   3.4 Find this proportion on the horizontal axis.
   3.5 Move vertically up from this point to find upper and lower bound on this estimate for population proportion.
   3.6 If this confidence interval is acceptable you have reached a satisfactory conclusion. If it is not acceptable increase your sample size until you have the required width of confidence interval.
   3.7 Draw this larger sample. Go back to 3.3 and repeat process until a satisfactory conclusion is reached.
   3.8 If the given confidence interval does not contain the proportion which the auditor accepts as satisfactory the auditor will reject the population.

Figures 21.1.1. to 21.1.3. are one tailed estimates. They tell us the probability of the population proportion exceeding the upper bound of the confidence interval.

Figures 21.2.1. to 21.2.3. are two tailed estimates. They tell us the probability of the population proportion being above the upper *and* below the lower bound.

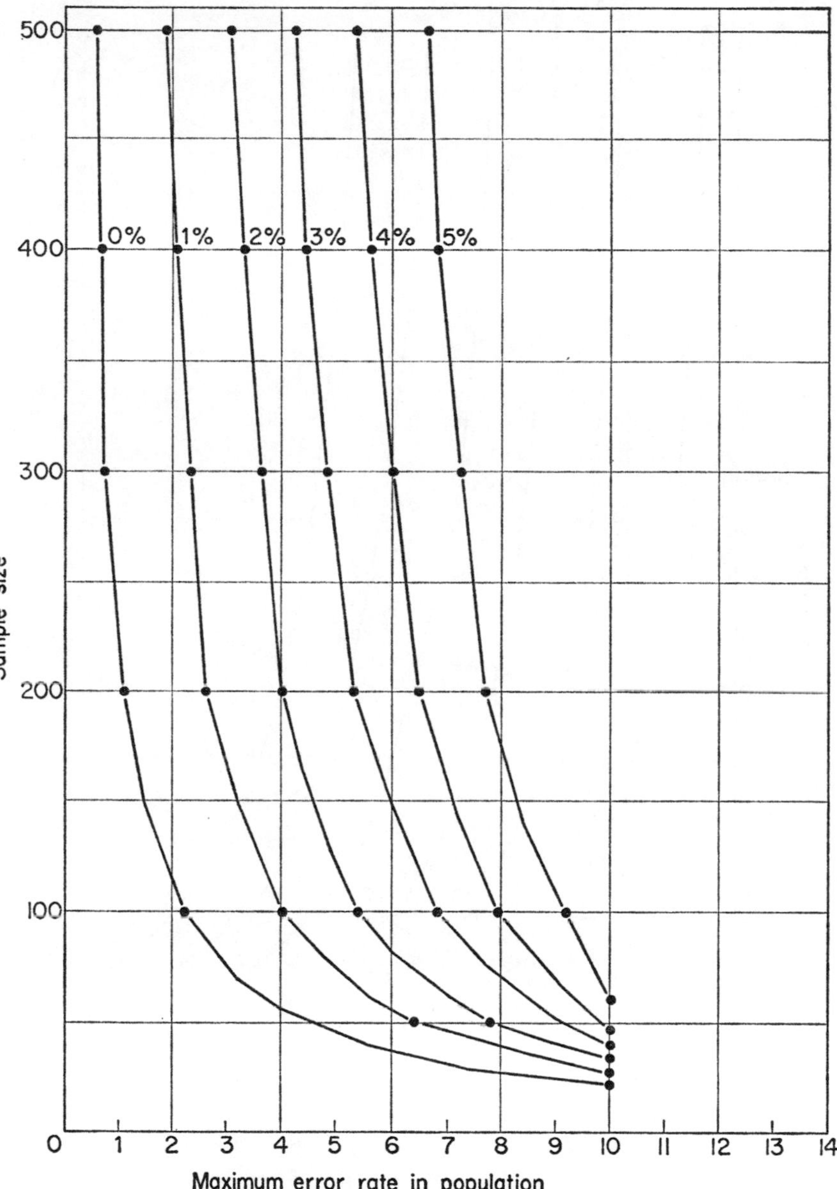


Fig. 21.1.1. Estimation sampling of error rates and other proportions.

| | |
|---|---|
| Population | 10,000 and over |
| Sample proportion having condition | 0% to 5% |
| Level of confidence | 90% |

The auditor is 90% confident that if the sample proportion is $x\%$ then the population proportion does not exceed $(x + y)\%$.

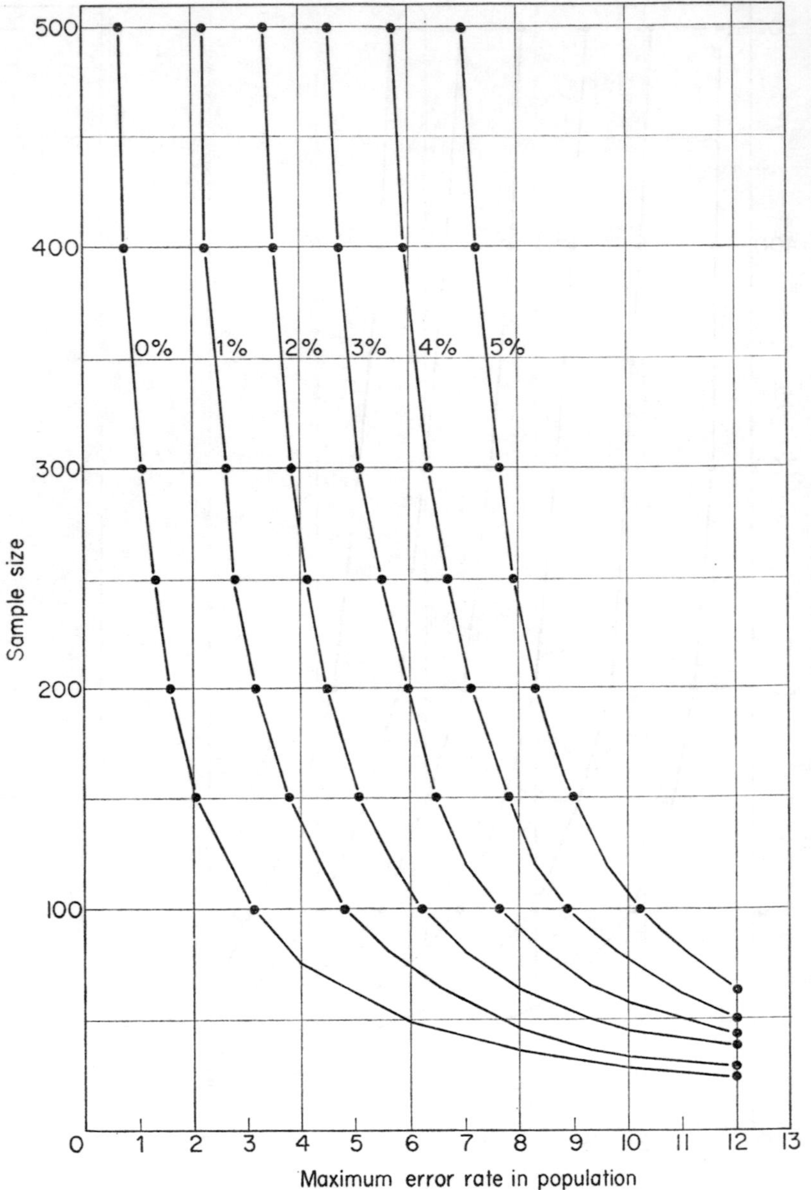


Fig. 21.1.2. Estimation sampling of error rates and other proportions.

| | |
|---|---|
| Population | 10,000 and over |
| Sample proportion having condition | 0% to 5% |
| Level of confidence | 95% |

The auditor is 95% confident that if the sample proportion is $x$% the population proportion does not exceed $(x + y)$%.

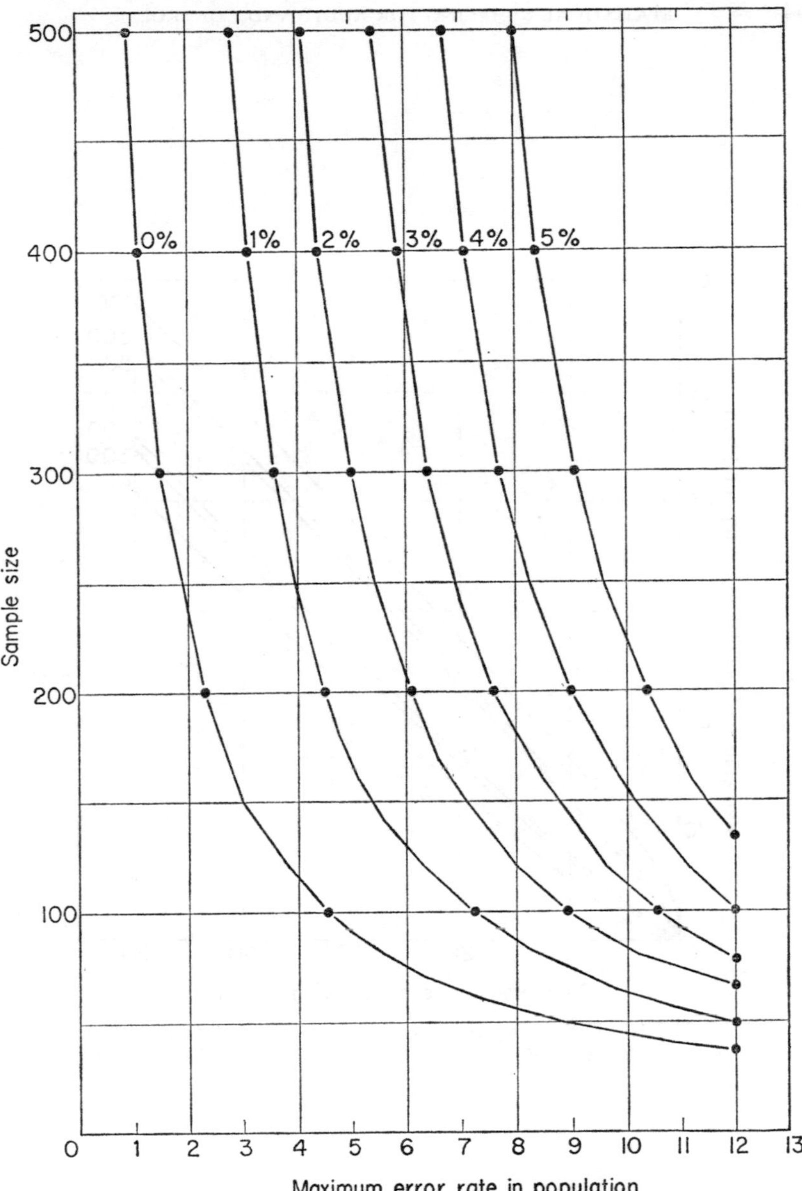


Fig. 21.1.3. Estimation sampling of error rates and other proportions.

| | |
|---|---|
| Population | 10,000 and over |
| Sample proportion having condition | 0% to 5% |
| Level of confidence | 99% |

The auditor is 99% confident that if the sample proportion is $x$% the population proportion does not exceed $(x + y)$%.

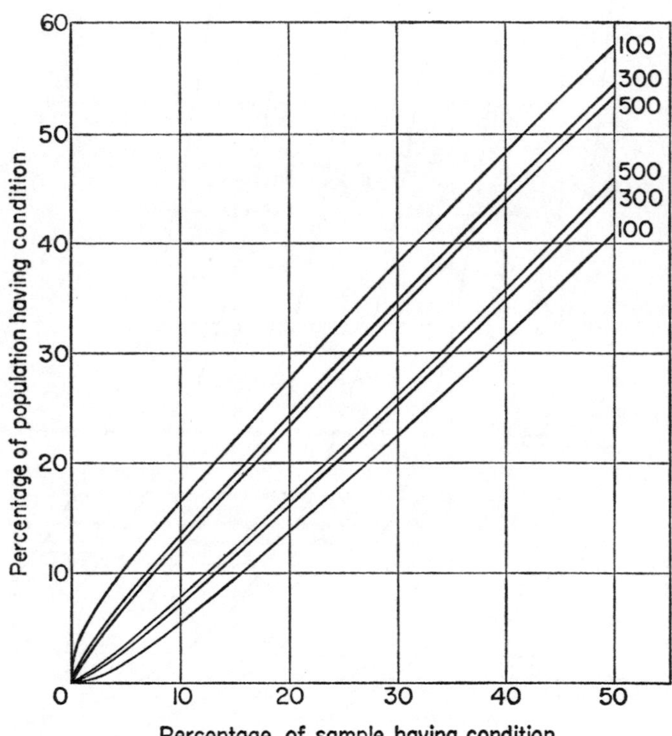


Fig. 21.2.1. Estimating the population proportion from a fixed sample size. Confidence level 90%. Population 10,000 and over.

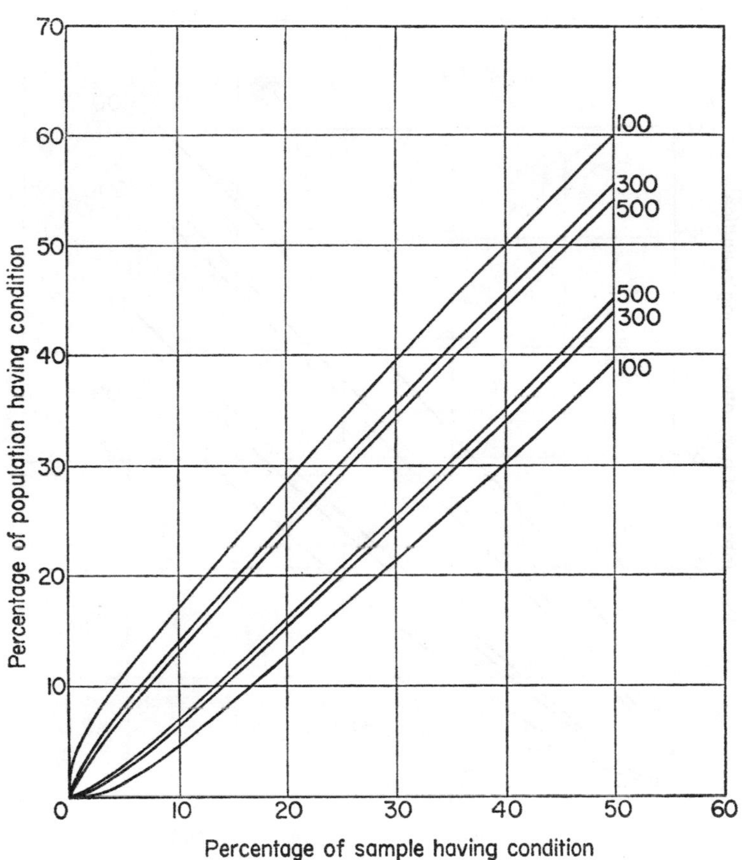


Fig. 21.2.2. Estimating the population proportion from a fixed sample size. Confidence level 95%. Population 10,000 and over.

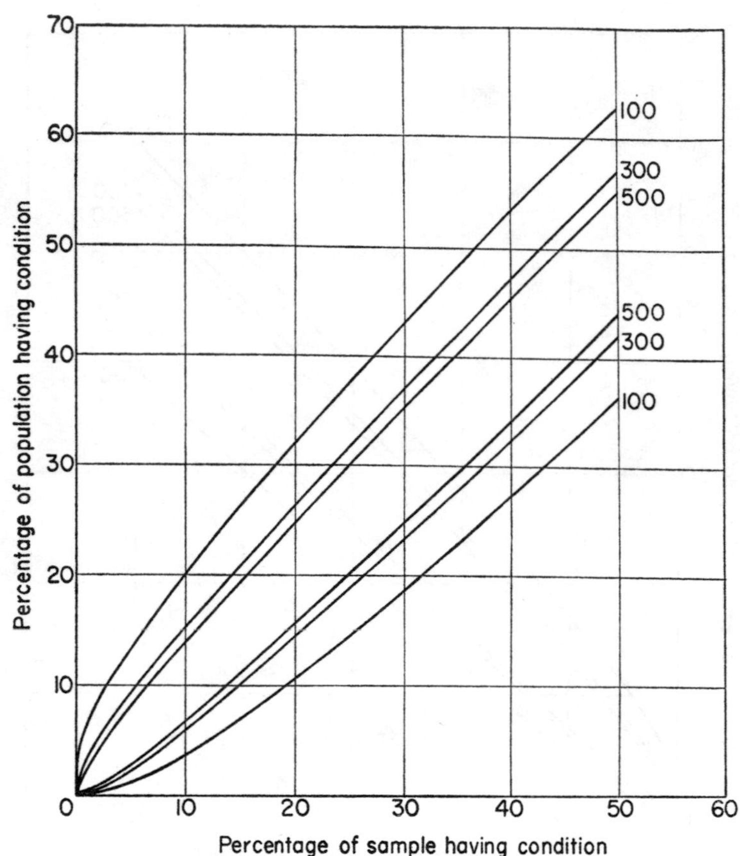


Fig. 21.2.3. Estimating the population proportion from a fixed sample size. Confidence level 99%. Population 10,000 and over.

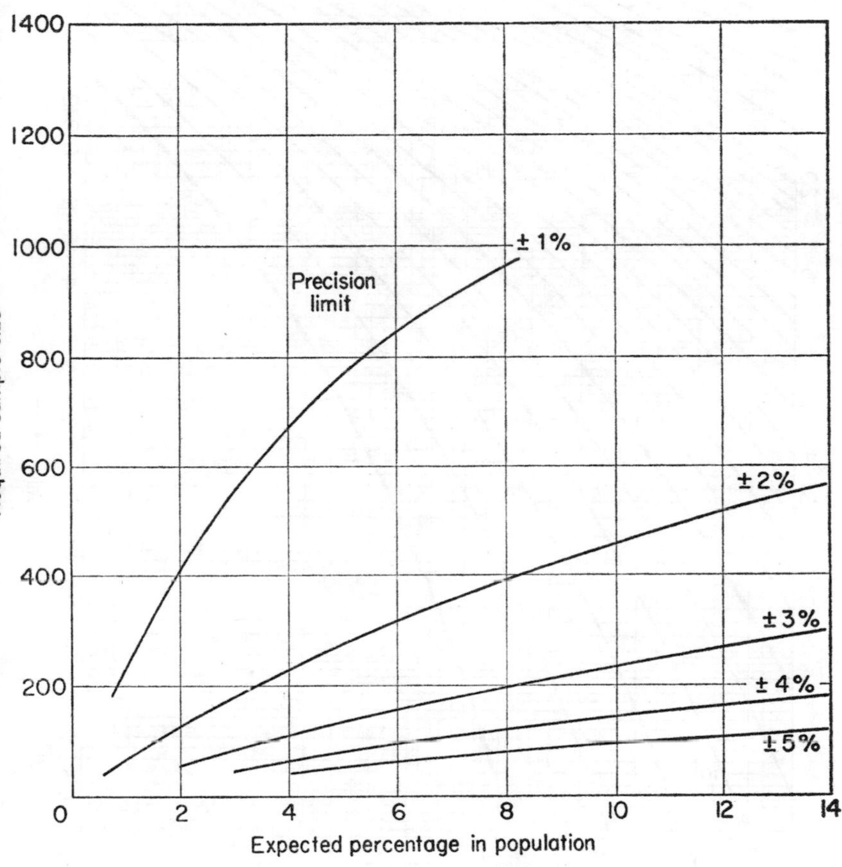


Fig. 21.3. Estimation sampling of attributes.

Population 2000

Confidence level 90%

The graph provides the required sample size given the expected proportion having condition and the required precision limit. Both the population size and the confidence level are fixed.

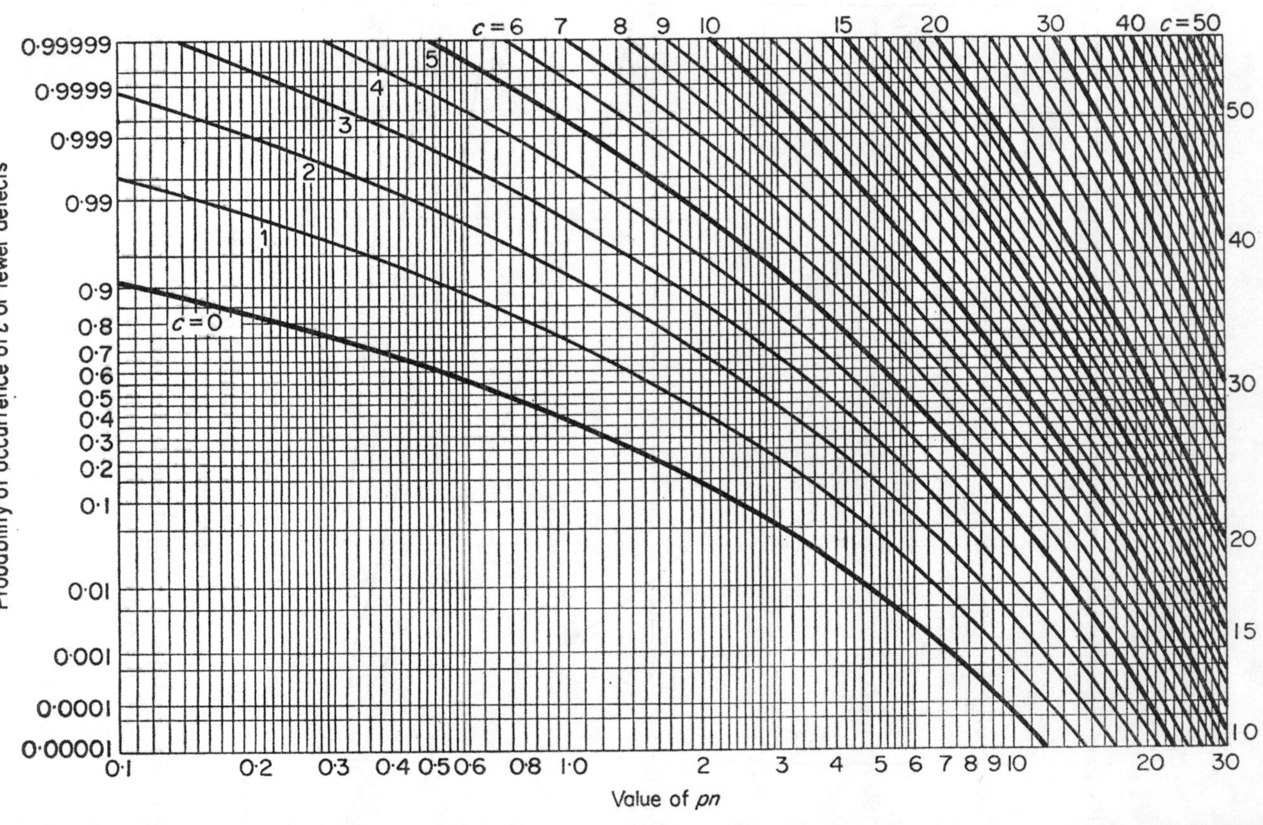
Probability of occurrence of c or fewer defects
0·99999
0·9999
0·999
0·99
0·9
0·8
0·7
0·6
0·5
0·4
0·3
0·2
0·1
0·01
0·001
0·0001
0·00001
c=0
1
2
3
4
5
c=6
7
8
9
10
15
20
30
40
c=50
0·1
0·2
0·3
0·4
0·5
0·6
0·8
1·0
2
3
4
5
6
7
8
9
10
20
30
Value of pn
50
40
30
20
15
10

## ESTIMATION SAMPLING OF VARIABLES (VALUES)

1. Decide on:

    Level of confidence.
    Precision limit (unit).

    Calculate:

    Population size.
    Standard deviation of population.

2. Calculate the ratio

    Unit precision limit/Standard deviation = $r$

3. Find $r$ on horizontal axis and move vertically up until you come to required level of confidence line.
4. Move across to horizontal axis to find required sample size.
5. Draw random sample of this size and calculate the mean value of this sample. Multiply this figure by the number of units in the population to arrive at an estimate of the population value.
6. Multiply the unit precision limit by the number of units in the population to arrive at the confidence interval on this estimate.
7. We can now state that we are $c\%$ confident that the actual population value lies between £$x \pm y$.

---

Fig. 21.4. Graph for estimating the probability of the population proportion being $x\%$ if the sample proportion is $y\%$.

$n$ = number of units in the sample
$p$ = percentage of units in the population having condition
$c$ = number of units in the sample having condition

For example if $n$ is 500 and $p$ is 1% we can calculate the probability of $c$ units turning up in the sample of 500.

| $c$ | Cumulative probability |
|---|---|
| 0 | 0·75 |
| 1 | 4·00 |
| 2 | 14·00 |
| 3 | 27·00 |

*Source*: By permission, from *Sampling Inspection Tables*, by Dodge and Romig, copyright 1944, John Wiley and Sons, Inc., p. 44.

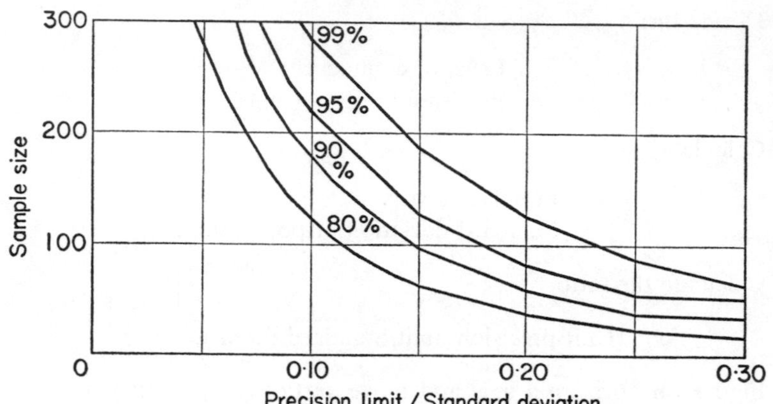


Fig. 21.5.1. Estimation sampling of variables.
Population 500
Graph shows required sample sizes given precision limit, standard deviation and required level of confidence.

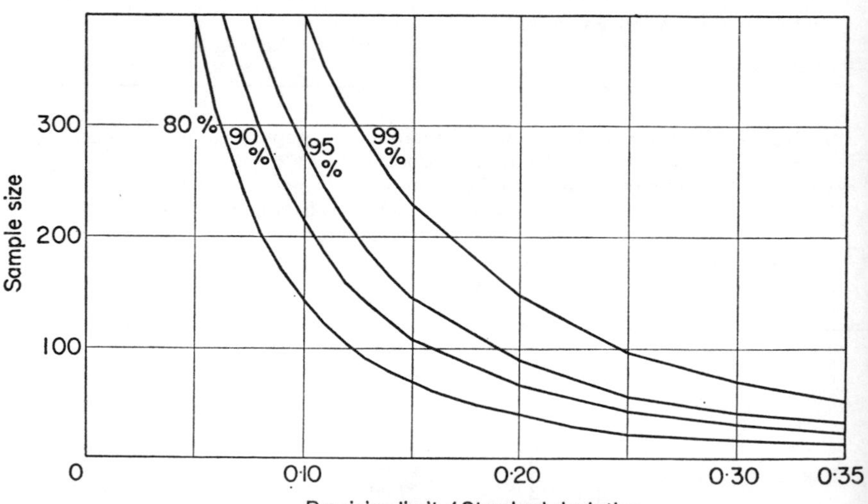


Fig. 21.5.2. Estimation sampling of variables.
Population 1000

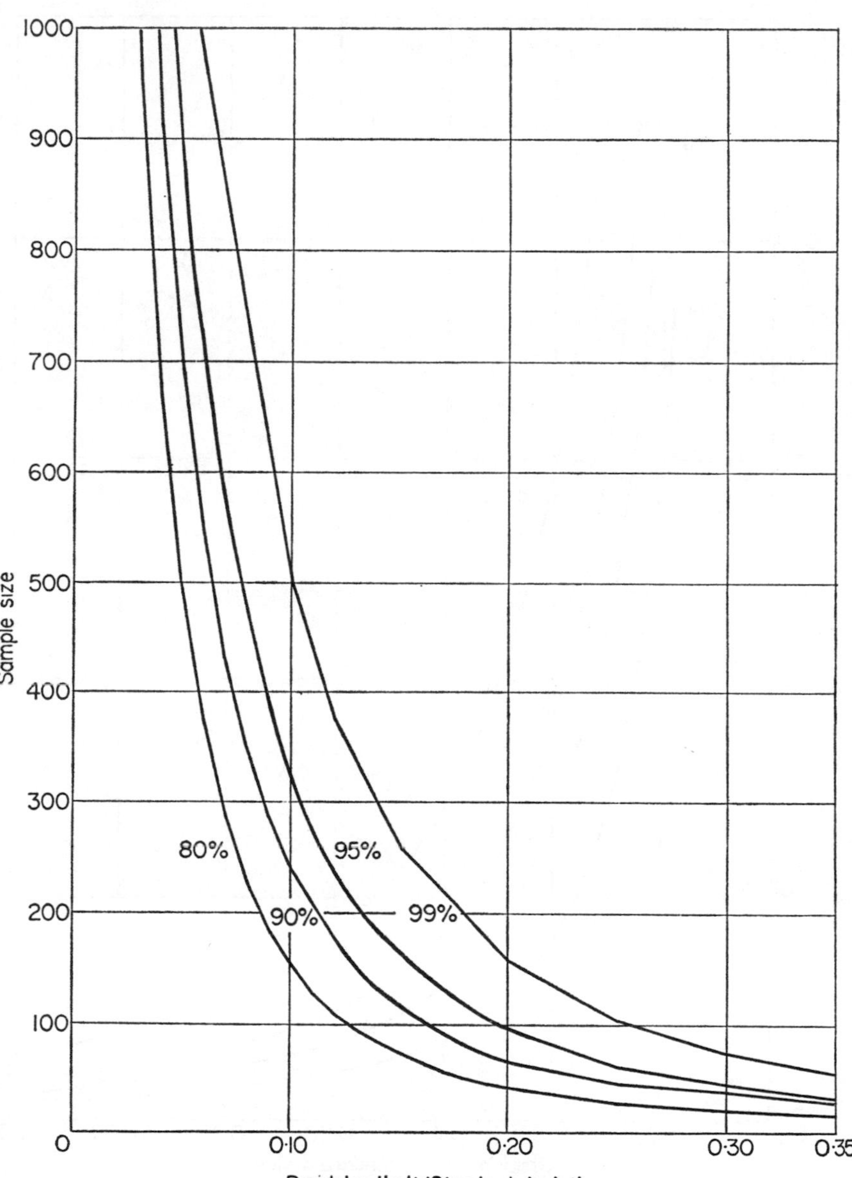


Fig. 21.5.3. Estimation sampling of variables.
Population 2000

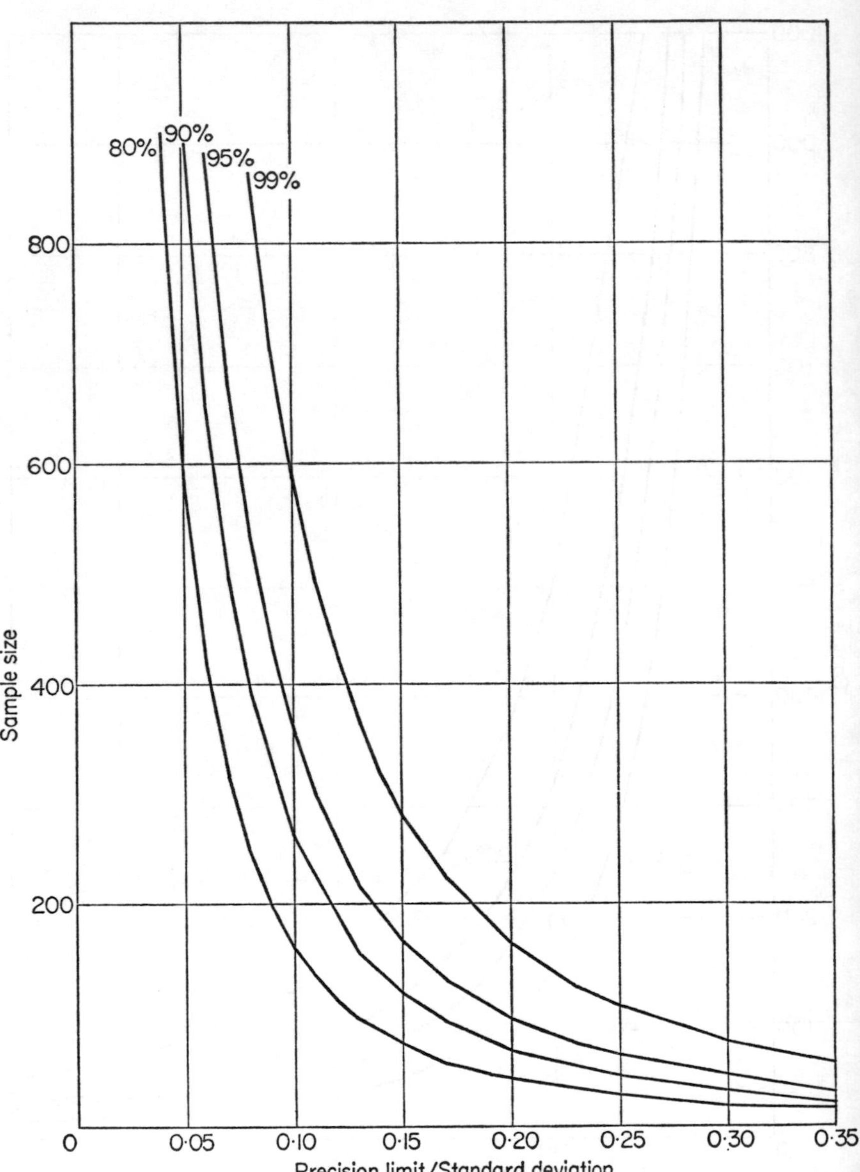


Fig. 21.5.4. Estimation sampling of variables.
Population 5000

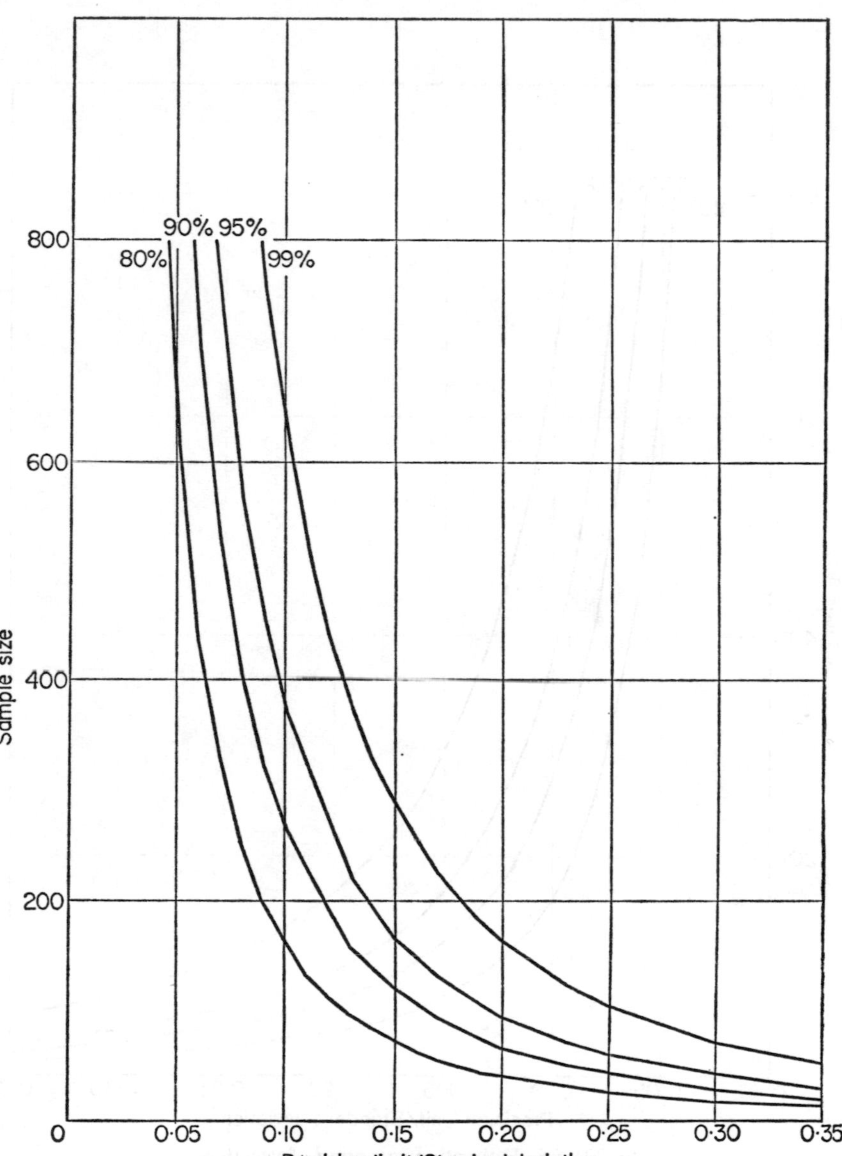


Fig. 21.5.5. Estimation sampling of variables.
Population 10,000

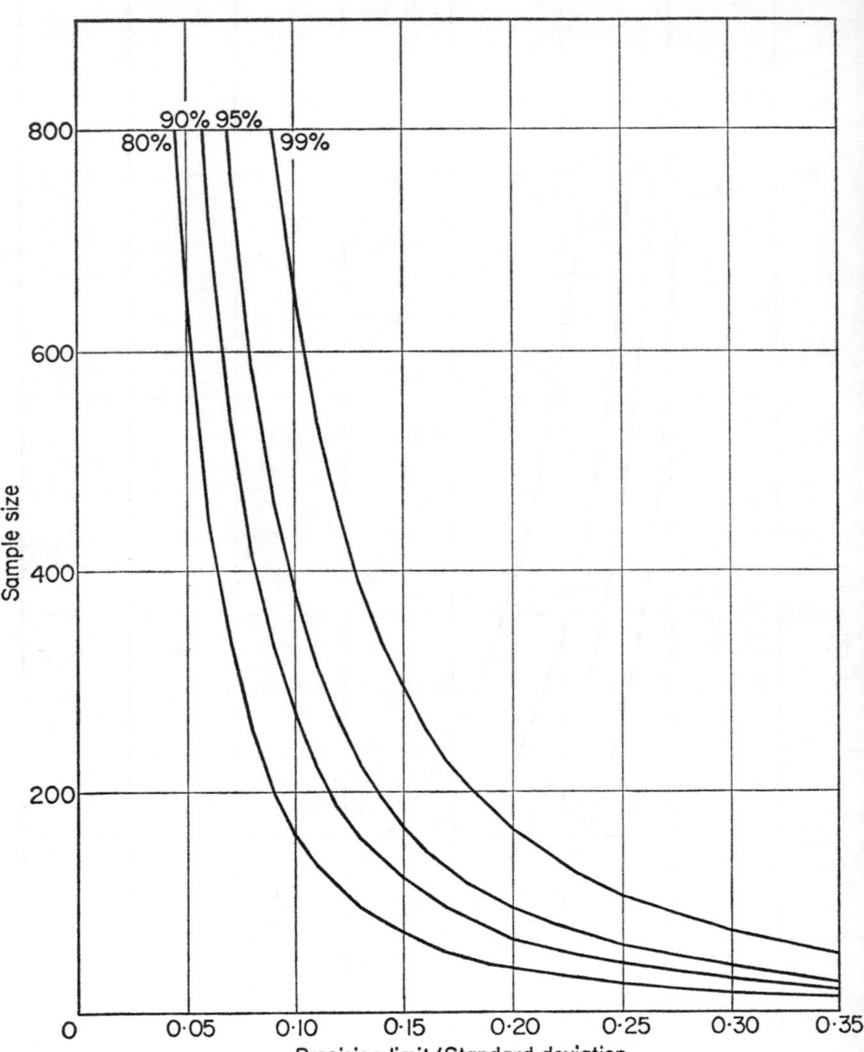


Fig. 21.5.6. Estimation sampling of variables.
Population 20,000

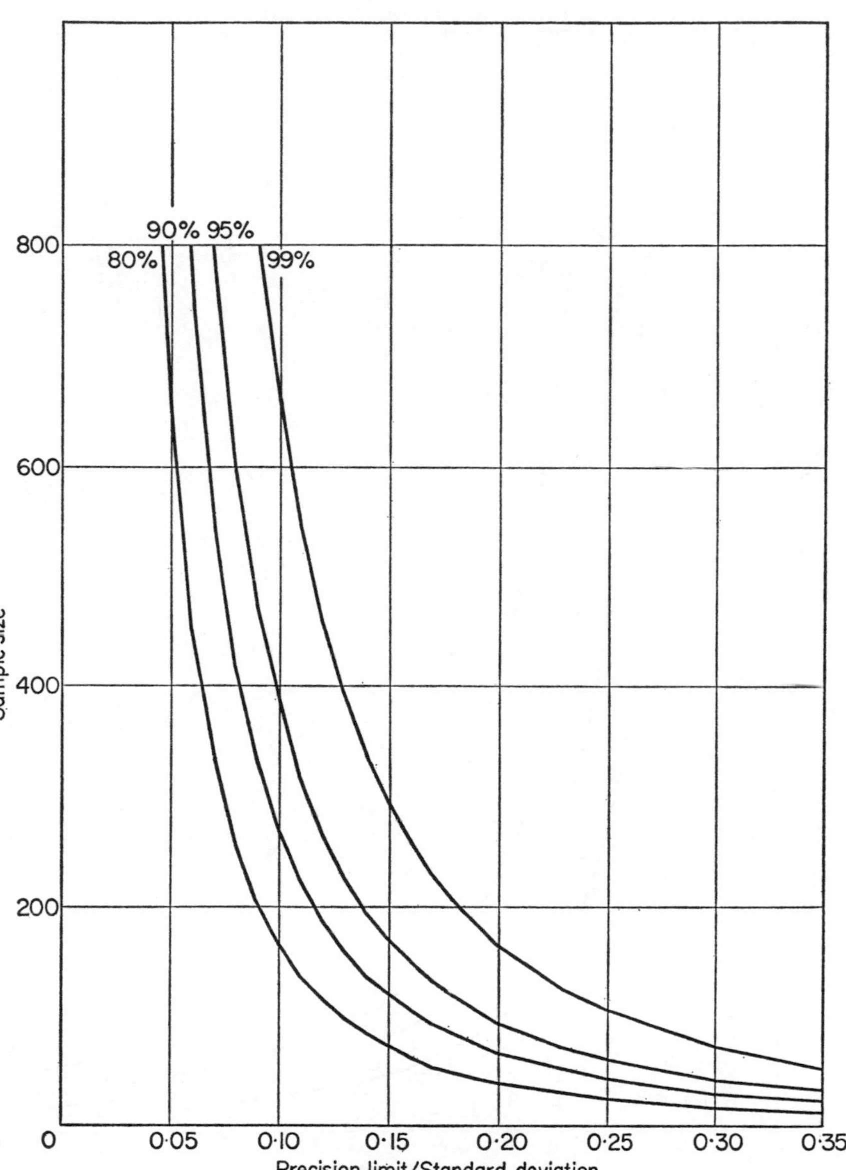


Fig. 21.5.7. Estimation sampling of variables.
Population 50,000

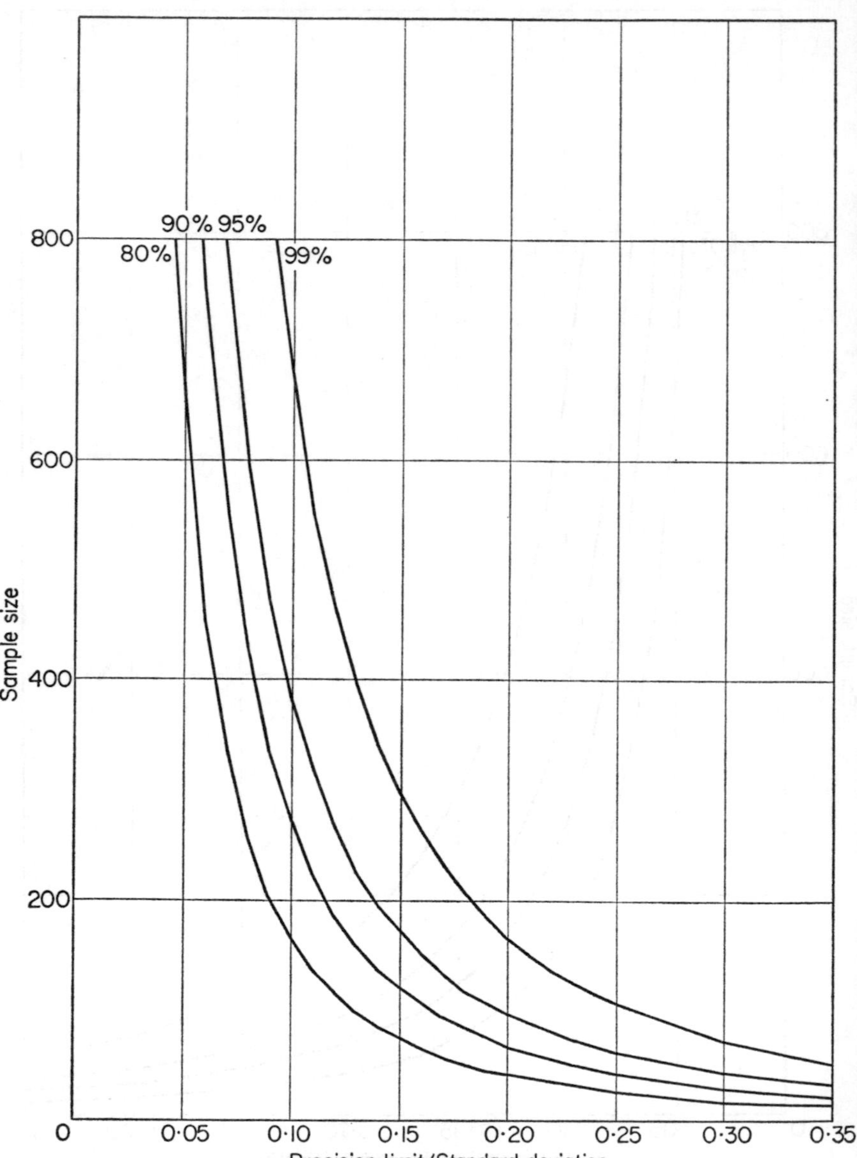


Fig. 21.5.8. Estimation sampling of variables.
Population infinite

## ACCEPTANCE SAMPLING

Using the tables.

1. Decide on:

    Minimum unacceptable error rate (MUER).
    Maximum unacceptable rejection rate (MURR).
    Level of confidence required in the MUER
    Level of confidence required in the MURR.

   Calculate:

    Population size (batch size).

2. Go to table of given population (batch) size.
3. Find row for MUER and MURR.
4. Move down both columns simultaneously until confidence levels close to those required for the MUER and the MURR are found.
5. Move horizontally along this row until you come to the acceptance number and the required sample size.
6. Draw a random sample of this size from batch.
7. If number of units with condition is equal to or less than the acceptable number accept the batch, otherwise reject the batch.

Table 21.3.1. Acceptance sampling. MURR or MUER. Population (batch) size 500. The figure within the matrix tells us the probability of accepting a batch containing a given error rate given; population size, sample size, maximum acceptable number of errors.

| *s* Sample size | *d* Acceptance number | When population contains error rate of: 0·2% | 0·4% | 1% | 2% | 3% | 10% |
|---|---|---|---|---|---|---|---|
| | | the probability of acceptance is | | | | | |
| 25 | 0 | 95·0 | 90·2 | 77·3 | 59·6 | 45·8 | 6·7 |
| 25 | 1 | 100·0 | 99·8 | 97·8 | 91·6 | 83·1 | 26·4 |
| 25 | 2 | | 100·0 | 99·9 | 99·0 | 96·6 | 53·4 |
| 25 | 3 | | | 99·9+ | 99·9 | 99·5 | 76·7 |
| 40 | 0 | 92·0 | 84·6 | 65·8 | 43·1 | 28·1 | 1·2 |
| 40 | 1 | 100·0 | 99·4 | 94·6 | 81·3 | 65·9 | 7·2 |
| 40 | 2 | | 100·0 | 99·6 | 96·2 | 89·0 | 21·1 |
| 40 | 3 | | | 99·9+ | 99·5 | 97·5 | 41·5 |
| 50 | 0 | 90·0 | 81·0 | 58·9 | 34·5 | 20·1 | 0·4 |
| 50 | 1 | 100·0 | 99·0 | 91·9 | 73·6 | 54·7 | 2·8 |
| 50 | 2 | | 100·0 | 99·2 | 93·2 | 81·8 | 9·9 |
| 50 | 3 | | | 99·9+ | 98·8 | 94·7 | 23·6 |
| 60 | 0 | 88·0 | 77·4 | 52·6 | 27·5 | 14·3 | 0·1 |
| 60 | 1 | 100·0 | 98·6 | 88·8 | 65·8 | 44·4 | 1·0 |
| 60 | 2 | | 100·0 | 98·6 | 89·3 | 73·6 | 4·3 |
| 60 | 3 | | | 99·9 | 97·7 | 90·7 | 12·1 |
| 75 | 0 | 85·0 | 72·2 | 44·2 | 19·4 | 8·4 | |
| 75 | 1 | 100·0 | 97·8 | 83·6 | 54·3 | 31·4 | 0·2 |
| 75 | 2 | | 100·0 | 97·4 | 82·2 | 60·4 | 1·1 |
| 75 | 3 | | | 99·8 | 95·2 | 82·6 | 3·8 |
| 100 | 0 | 80·0 | 64·0 | 32·6 | 10·5 | 3·3 | |
| 100 | 1 | 100·0 | 96·0 | 73·8 | 37·3 | 16·3 | |
| 100 | 2 | | 100·0 | 94·3 | 67·8 | 39·5 | 0·1 |
| 100 | 3 | | | 99·4 | 88·1 | 64·9 | 0·4 |
| 100 | 4 | | | 99·9+ | 96·9 | 83·9 | 1·4 |
| 125 | 0 | 75·0 | 56·2 | 23·6 | 5·5 | 1·2 | |
| 125 | 1 | 100·0 | 93·8 | 63·8 | 24·2 | 7·7 | |
| 125 | 2 | | 100·0 | 89·8 | 52·5 | 24·3 | |
| 125 | 3 | | | 98·5 | 77·8 | 48·7 | |
| 125 | 4 | | | 99·9 | 92·4 | 73·2 | 0·1 |
| 150 | 0 | 70·0 | 49·0 | 16·7 | 2·7 | 0·4 | |
| 150 | 1 | 100·0 | 91·0 | 52·8 | 14·7 | 3·3 | |
| 150 | 2 | | 100·0 | 83·8 | 38·1 | 12·3 | |
| 150 | 3 | | | 97·0 | 65·0 | 29·3 | |
| 150 | 4 | | | 99·8 | 85·2 | 51·5 | |
| 175 | 0 | 65·0 | 42·2 | 11·5 | 1·3 | 0·1 | |
| 175 | 1 | 100·0 | 87·8 | 42·8 | 8·4 | 1·3 | |
| 175 | 2 | | 100·0 | 76·6 | 25·9 | 5·9 | |
| 175 | 3 | | | 94·7 | 51·3 | 16·9 | |
| 175 | 4 | | | 99·5 | 75·3 | 34·9 | |

Table 21.3.2. Acceptance sampling. MURR or MUER. Population (batch) size 1000.

| s Sample size | d Acceptance number | When population contains error rate of: 0·5% | 1% | 2% | 3% | 10% |
|---|---|---|---|---|---|---|
| | | the probability of accepting it is | | | | |
| 25 | 0 | 88·1 | 77·5 | 60·0 | 46·3 | 6·9 |
| 25 | 1 | 99·4 | 97·6 | 91·3 | 82·9 | 26·7 |
| 25 | 2 | 99·9+ | 99·9 | 98·8 | 96·4 | 53·6 |
| 25 | 3 | 99·9+ | 99·9+ | 99·9 | 99·5 | 77·0 |
| 50 | 0 | 77·3 | 59·7 | 35·5 | 21·0 | 0·4 |
| 50 | 1 | 97·8 | 91·5 | 73·6 | 55·1 | 3·1 |
| 50 | 2 | 99·9 | 98·9 | 92·6 | 81·4 | 10·6 |
| 50 | 3 | 99·9+ | 99·9 | 98·5 | 94·2 | 24·3 |
| 60 | 0 | 73·3 | 53·7 | 28·7 | 15·2 | 0·1 |
| 60 | 1 | 96·9 | 88·3 | 66·0 | 45·2 | 1·2 |
| 60 | 2 | 99·8 | 98·2 | 88·7 | 73·3 | 4·8 |
| 60 | 3 | 99·9+ | 99·8 | 97·2 | 90·0 | 12·9 |
| 75 | 0 | 67·7 | 45·7 | 20·7 | 9·3 | |
| 75 | 1 | 95·2 | 83·1 | 55·0 | 32·7 | 0·3 |
| 75 | 2 | 99·6 | 96·7 | 81·6 | 60·6 | 1·3 |
| 75 | 3 | 99·9+ | 99·6 | 94·4 | 81·8 | 4·4 |
| 100 | 0 | 59·0 | 34·7 | 11·9 | 4·0 | |
| 100 | 1 | 91·9 | 73·6 | 38·9 | 17·9 | |
| 100 | 2 | 99·2 | 93·1 | 67·7 | 40·8 | 0·1 |
| 100 | 3 | 99·9+ | 98·8 | 86·9 | 64·8 | 0·6 |
| 100 | 4 | 99·9+ | 99·8 | 95·8 | 82·7 | 1·9 |
| 125 | 0 | 51·2 | 26·1 | 7·3 | 1·7 | |
| 125 | 1 | 88·0 | 68·9 | 26·4 | 9·3 | |
| 125 | 2 | 98·4 | 88·2 | 53·4 | 25·4 | |
| 125 | 3 | 99·9 | 97·3 | 76·7 | 47·1 | 0·1 |
| 125 | 4 | 99·9+ | 99·6 | 90·7 | 68·2 | 0·2 |
| 150 | 0 | 44·3 | 19·5 | 3·7 | 0·7 | |
| 150 | 1 | 83·6 | 54·4 | 17·3 | 4·6 | |
| 150 | 2 | 97·4 | 82·1 | 40·3 | 14·7 | |
| 150 | 3 | 99·8 | 95·1 | 64·8 | 31·8 | |
| 150 | 4 | 99·9+ | 99·0 | 83·2 | 52·3 | |
| 200 | 0 | 32·7 | 10·6 | 1·1 | 0·1 | |
| 200 | 1 | 73·8 | 37·6 | 6·7 | 1·0 | |
| 200 | 2 | 94·3 | 68·3 | 20·3 | 4·2 | |
| 200 | 3 | 99·3 | 88·2 | 41·0 | 11·9 | |
| 200 | 4 | 99·9+ | 97·0 | 62·7 | 25·1 | |
| 200 | 5 | 100·0 | 99·6 | 80·4 | 42·5 | |
| 225 | 0 | 27·9 | 7·7 | 0·6 | 0·1 | |
| 225 | 1 | 68·6 | 30·1 | 4·0 | 0·4 | |
| 225 | 2 | 92·2 | 59·5 | 13·7 | 2·1 | |
| 225 | 3 | 99·0 | 82·2 | 30·8 | 6·7 | |
| 225 | 4 | 99·9 | 93·7 | 52·0 | 15·9 | |
| 225 | 5 | 100·0 | 97·7 | 71·7 | 29·9 | |

Table 21.3.2—*Continued*

| *s* Sample size | *d* Acceptance number | When population contains error rate of: 0·5% | 1% | 2% | 3% | 10% |
|---|---|---|---|---|---|---|
| | | the probability of accepting it is | | | | |
| 275 | 0 | 20·0 | 3·9 | 0·1 | 0·1 | |
| 275 | 1 | 58·0 | 19·1 | 1·3 | 0·1 | |
| 275 | 2 | 86·9 | 45·1 | 5·6 | 0·4 | |
| 275 | 3 | 97·8 | 71·6 | 15·5 | 1·8 | |
| 275 | 4 | 99·9 | 89·0 | 31·7 | 3·1 | |
| 275 | 5 | 100·0 | 96·9 | 51·5 | 5·8 | |
| 300 | 0 | 16·7 | 2·8 | 0·1 | 0·1 | |
| 300 | 1 | 52·8 | 14·8 | 0·7 | 0·1 | |
| 300 | 2 | 83·8 | 38·2 | 3·4 | 0·2 | |
| 300 | 3 | 97·0 | 65·0 | 10·5 | 0·8 | |
| 300 | 4 | 99·8 | 85·1 | 23·5 | 2·8 | |
| 300 | 5 | 100·0 | 95·3 | 41·5 | 7·3 | |

Adapted, by permission, from Arkin (1963)

Table 21.3.3. Acceptance sampling. MURR or MUER. Population (batch) size 2000.

| *s* Sample size | *d* Acceptance number | When population contains error rate of: 0·5% | 1% | 2% | 3% | 10% |
|---|---|---|---|---|---|---|
| | | the probability of accepting it is | | | | |
| 50 | 0 | 77·6 | 60·1 | 36·0 | 21·4 | 0·5 |
| 50 | 1 | 97·6 | 91·3 | 73·6 | 55·3 | 3·2 |
| 50 | 2 | 99·8 | 98·8 | 92·4 | 81·3 | 10·9 |
| 50 | 3 | 99·9+ | 99·9 | 98·4 | 94·0 | 24·7 |
| 60 | 0 | 73·7 | 54·2 | 29·2 | 15·6 | 0·2 |
| 60 | 1 | 96·6 | 88·1 | 66·1 | 45·6 | 1·3 |
| 60 | 2 | 99·7 | 98·0 | 88·4 | 73·3 | 5·1 |
| 60 | 3 | 99·9+ | 99·8 | 97·0 | 89·7 | 13·4 |
| 75 | 0 | 68·2 | 46·4 | 21·3 | 9·7 | |
| 75 | 1 | 94·9 | 82·9 | 55·3 | 33·2 | 0·3 |
| 75 | 2 | 99·5 | 96·4 | 81·3 | 60·7 | 1·5 |
| 75 | 3 | 99·9+ | 99·4 | 94·0 | 81·5 | 4·8 |
| 90 | 0 | 63·0 | 39·6 | 15·6 | 6·0 | |
| 90 | 1 | 92·9 | 77·4 | 45·5 | 23·7 | 0·1 |
| 90 | 2 | 99·2 | 94·2 | 73·3 | 48·7 | 0·4 |
| 90 | 3 | 99·9 | 98·9 | 89·8 | 71·7 | 1·5 |
| 100 | 0 | 59·8 | 35·7 | 12·6 | 4·4 | |
| 100 | 1 | 91·4 | 73·6 | 39·6 | 18·7 | |
| 100 | 2 | 98·9 | 92·6 | 67·7 | 41·4 | 0·2 |
| 100 | 3 | 99·9 | 97·9 | 94·1 | 64·8 | 0·7 |
| 100 | 4 | 99·9+ | 99·0 | 98·5 | 82·2 | 2·1 |

Table 21.3.3—*Continued*

| *s* Sample size | *d* Acceptance number | When population contains error rate of: 0·5% | 1% | 2% | 3% | 10% |
|---|---|---|---|---|---|---|
| | | the probability of accepting it is | | | | |
| 125 | 0 | 52·4 | 27·3 | 7·4 | 2·0 | |
| 125 | 1 | 87·5 | 64·2 | 27·4 | 10·1 | |
| 125 | 2 | 97·9 | 87·5 | 53·9 | 26·3 | |
| 125 | 3 | 99·9 | 96·8 | 76·2 | 47·7 | 0·1 |
| 125 | 4 | 99·9+ | 99·4 | 90·0 | 68·0 | 0·3 |
| 150 | 0 | 45·8 | 20·9 | 4·3 | 0·9 | |
| 150 | 1 | 83·1 | 55·1 | 18·5 | 5·2 | |
| 150 | 2 | 96·7 | 81·5 | 41·2 | 15·9 | |
| 150 | 3 | 99·6 | 94·3 | 64·8 | 32·9 | |
| 150 | 4 | 99·9+ | 98·6 | 82·4 | 52·7 | |
| 200 | 0 | 34·8 | 12·0 | 1·4 | 0·2 | |
| 200 | 1 | 73·6 | 39·1 | 7·8 | 1·3 | |
| 200 | 2 | 93·0 | 67·7 | 22·0 | 5·1 | |
| 200 | 3 | 98·7 | 86·8 | 42·1 | 13·4 | |
| 200 | 4 | 99·8 | 95·8 | 62·9 | 26·7 | |
| 200 | 5 | 99·9+ | 98·9 | 79·6 | 43·5 | |
| 300 | 0 | 19·6 | 3·8 | 0·1 | | |
| 300 | 1 | 54·4 | 17·4 | 1·2 | 0·1 | |
| 300 | 2 | 82·1 | 40·4 | 4·7 | 0·4 | |
| 300 | 3 | 95·1 | 64·8 | 12·8 | 1·4 | |
| 300 | 4 | 99·0 | 83·1 | 26·1 | 4·0 | |
| 300 | 5 | 99·8 | 93·4 | 43·1 | 9·3 | |
| 400 | 0 | 10·7 | 1·1 | 0·1 | | |
| 400 | 1 | 37·5 | 6·8 | 0·1 | | |
| 400 | 2 | 67·8 | 20·5 | 0·8 | | |
| 400 | 3 | 88·0 | 41·1 | 2·7 | | |
| 400 | 4 | 96·8 | 63·0 | 7·4 | | |
| 400 | 5 | 99·4 | 80·5 | 15·9 | 0·1 | |

Adapted, by permission, from Arkin (1963)

## METHOD OF CONSTRUCTING OTHER SINGLE SAMPLING PLANS

J.M. Cameron has constructed a table from which auditors can design their own acceptance sampling plan. This is provided in Table 21.4.

Before using the table the auditor must decide on a suitable MUER and MURR. Cameron provides six 'levels of confidence' schemes for the auditor

| | % | % | % |
|---|---|---|---|
| Level of confidence in MUER (*B*) | 90 | 95 | 99 |
| Level of confidence in MURR (*d*) | 95 | 95 | 95 |
| MUER (*B*) | 90 | 95 | 99 |
| MURR (*d*) | 99 | 99 | 99 |

Table 21.4. Cameron's table for calculating single acceptance sampling plans.

| | MUER/MURR | | | | | MUER/MURR | | | |
|---|---|---|---|---|---|---|---|---|---|
| $c$ | $d = 0{\cdot}05$ $B = 0{\cdot}10$ | $d = 0{\cdot}05$ $B = 0{\cdot}05$ | $d = 0{\cdot}05$ $B = 0{\cdot}01$ | $x$ | $c$ | $d = 0{\cdot}01$ $B = 0{\cdot}10$ | $d = 0{\cdot}01$ $B = 0{\cdot}05$ | $d = 0{\cdot}01$ $B = 0{\cdot}01$ | $x$ |
| 0 | 44·890 | 58·404 | 89·781 | 0·052 | 0 | 229·105 | 298·073 | 458·210 | 0·010 |
| 1 | 10·946 | 13·349 | 18·681 | 0·355 | 1 | 26·184 | 31·933 | 44·686 | 0·149 |
| 2 | 6·509 | 7·699 | 10·280 | 0·818 | 2 | 12·206 | 14·439 | 19·278 | 0·436 |
| 3 | 4·890 | 5·675 | 7·352 | 1·366 | 3 | 8·115 | 9·418 | 12·202 | 0·823 |
| 4 | 4·057 | 4·646 | 5·890 | 1·970 | 4 | 6·249 | 7·156 | 9·072 | 1·279 |
| 5 | 3·549 | 4·023 | 5·017 | 2·613 | 5 | 5·195 | 5·889 | 7·343 | 1·785 |
| 6 | 3·206 | 3·604 | 4·435 | 3·286 | 6 | 4·520 | 5·082 | 6·253 | 2·330 |
| 7 | 2·957 | 3·303 | 4·019 | 3·981 | 7 | 4·050 | 4·524 | 5·506 | 2·906 |
| 8 | 2·768 | 3·074 | 3·707 | 4·695 | 8 | 3·705 | 4·115 | 4·962 | 3·507 |
| 9 | 2·618 | 2·895 | 3·462 | 5·426 | 9 | 3·440 | 3·803 | 4·548 | 4·130 |
| 10 | 2·497 | 2·750 | 3·265 | 6·169 | 10 | 3·229 | 3·555 | 4·222 | 4·771 |
| 11 | 2·397 | 2·630 | 3·104 | 6·924 | 11 | 3·058 | 3·354 | 3·959 | 5·428 |
| 12 | 2·312 | 2·528 | 2·968 | 7·690 | 12 | 2·915 | 3·188 | 3·742 | 6·099 |
| 13 | 2·240 | 2·442 | 2·852 | 8·464 | 13 | 2·795 | 3·047 | 3·559 | 6·782 |
| 14 | 2·177 | 2·367 | 2·752 | 9·246 | 14 | 2·692 | 2·927 | 3·403 | 7·477 |
| 15 | 2·122 | 2·302 | 2·665 | 10·035 | 15 | 2·603 | 2·823 | 3·269 | 8·181 |
| 16 | 2·073 | 2·244 | 2·588 | 10·831 | 16 | 2·524 | 2·732 | 3·151 | 8·895 |
| 17 | 2·029 | 2·192 | 2·520 | 11·633 | 17 | 2·455 | 2·652 | 3·048 | 9·616 |
| 18 | 1·990 | 2·145 | 2·458 | 12·442 | 18 | 2·393 | 2·580 | 2·956 | 10·346 |
| 19 | 1·954 | 2·103 | 2·403 | 13·254 | 19 | 2·337 | 2·516 | 2·874 | 11·082 |

| | | | | | | | | | |
|---|---|---|---|---|---|---|---|---|---|
| 20 | 1·922 | 2·065 | 2·352 | 14·072 | 20 | 2·287 | 2·458 | 2·799 | 11·825 |
| 21 | 1·892 | 2·030 | 2·307 | 14·894 | 21 | 2·241 | 2·405 | 2·733 | 12·574 |
| 22 | 1·865 | 1·999 | 2·265 | 15·719 | 22 | 2·200 | 2·357 | 2·671 | 13·329 |
| 23 | 1·840 | 1·969 | 2·226 | 16·548 | 23 | 2·162 | 2·313 | 2·615 | 14·088 |
| 24 | 1·817 | 1·942 | 2·191 | 17·382 | 24 | 2·126 | 2·272 | 2·564 | 14·853 |
| 25 | 1·795 | 1·917 | 2·158 | 18·218 | 25 | 2·094 | 2·235 | 2·516 | 15·623 |
| 26 | 1·775 | 1·893 | 2·127 | 19·058 | 26 | 2·064 | 2·200 | 2·472 | 16·397 |
| 27 | 1·757 | 1·871 | 2·098 | 19·900 | 27 | 2·035 | 2·168 | 2·431 | 17·175 |
| 28 | 1·739 | 1·850 | 2·071 | 20·746 | 28 | 2·009 | 2·138 | 2·393 | 17·957 |
| 29 | 1·723 | 1·831 | 2·046 | 21·594 | 29 | 1·985 | 2·110 | 2·358 | 18·742 |
| 30 | 1·707 | 1·813 | 2·023 | 22·444 | 30 | 1·962 | 2·083 | 2·324 | 19·532 |
| 31 | 1·692 | 1·796 | 2·001 | 23·298 | 31 | 1·940 | 2·059 | 2·293 | 20·324 |
| 32 | 1·679 | 1·780 | 1·980 | 24·152 | 32 | 1·920 | 2·035 | 2·264 | 21·120 |
| 33 | 1·665 | 1·764 | 1·960 | 25·010 | 33 | 1·900 | 2·013 | 2·236 | 21·919 |
| 34 | 1·653 | 1·750 | 1·941 | 25·870 | 34 | 1·882 | 1·992 | 2·210 | 22·721 |
| 35 | 1·641 | 1·736 | 1·923 | 26·731 | 35 | 1·865 | 1·973 | 2·185 | 23·525 |
| 36 | 1·630 | 1·723 | 1·906 | 27·594 | 36 | 1·848 | 1·954 | 2·162 | 24·333 |
| 37 | 1·619 | 1·710 | 1·890 | 28·460 | 37 | 1·833 | 1·936 | 2·139 | 25·143 |
| 38 | 1·609 | 1·698 | 1·875 | 29·327 | 38 | 1·818 | 1·920 | 2·118 | 25·955 |
| 39 | 1·599 | 1·687 | 1·860 | 30·196 | 39 | 1·804 | 1·903 | 2·098 | 26·770 |
| 40 | 1·590 | 1·676 | 1·846 | 31·066 | 40 | 1·790 | 1·883 | 2·079 | 27·587 |
| 41 | 1·581 | 1·666 | 1·833 | 31·938 | 41 | 1·777 | 1·873 | 2·060 | 28·406 |
| 42 | 1·572 | 1·656 | 1·820 | 32·812 | 42 | 1·765 | 1·859 | 2·043 | 29·228 |
| 43 | 1·564 | 1·646 | 1·807 | 33·686 | 43 | 1·753 | 1·845 | 2·026 | 30·051 |
| 44 | 1·556 | 1·637 | 1·796 | 34·563 | 44 | 1·742 | 1·832 | 2·010 | 30·877 |
| 45 | 1·548 | 1·628 | 1·784 | 35·441 | 45 | 1·731 | 1·820 | 1·994 | 31·704 |
| 46 | 1·541 | 1·619 | 1·773 | 36·320 | 46 | 1·720 | 1·808 | 1·980 | 32·534 |
| 47 | 1·534 | 1·611 | 1·763 | 37·200 | 47 | 1·710 | 1·796 | 1·965 | 33·365 |
| 48 | 1·527 | 1·603 | 1·752 | 38·082 | 48 | 1·701 | 1·785 | 1·952 | 34·198 |
| 49 | 1·521 | 1·596 | 1·743 | 38·965 | 49 | 1·691 | 1·775 | 1·938 | 35·032 |

*Source*: By permission, from *Industrial Quality Control*, article by J. M. Cameron, July 1952, p. 39. 

The auditor uses the table as follows.

1. Calculate the ratio MUER/MURR.
2. Select that column which provides the required confidence levels in MUER and MURR.
3. Find MUER/MURR ratio in this column. If the exact figure is not available take the next higher figure.
4. The acceptance number is the figure in the $c$ column of that row.
5. The sample size is found by dividing the $x$ column by the MURR.

*Example.*

| | |
|---|---|
| Population | 1000 |
| MUER | 3% (90% CL) |
| MURR | 1% (95% CL) |
| MUER/MURR | 3 |
| $x$/MURR = 3·286/0·01 = | 329. |

A suitable sampling plan is (1000, 329, 6).

The sampling plans provided tend to be rather conservative approximations of precisely calculated tables since the batch is assumed to be very large, say over 10,000.

## INSTRUCTIONS FOR CALCULATING A DOUBLE SAMPLING PLAN

1. Calculate MUER/MURR = $R$.
2. Find $R$ in the first column, or next higher value.
3. Read off

$c$ = acceptance number for the first sample
$k$ = rejection number for the first sample
$d$ = acceptance number for the double sample

4. The rejection number for the double sample is found by adding 1 to the acceptance number for the double sample.
5. The plan is designed so that both samples are of equal size.

To calculate this sample size we divide the figure in the last column by the MURR.

*Example.*

1. MUER = 3%, MURR = 1%, MUER/MURR = 3·00.
2. Find 3·0 in the first column.
3. $c = 3$, $k = 7$, $d = 9$.
4. The rejection number for the double sample is $9 + 1 = 10$.
5. The sample size is 2·77/0·01 = 277.

Thus the final sampling plan is,

First sample 277. Accept on 3, reject on 7.
Second sample 554. Accept on 9, reject on 10.

Table 21.5. For calculating double sampling plan. (The level of confidence for both MUER and MURR is 95%.)

| MURR | $c_1$ | $k_1$ | $d$ | $n_1p_1$ |
|---|---|---|---|---|
| 15·1 | 0 | 1 | 1 | 0·207 |
| 8·3 | 0 | 2 | 2 | 0·427 |
| 5·1 | 1 | 3 | 4 | 1·00 |
| 4·1 | 2 | 4 | 6 | 1·63 |
| 3·5 | 2 | 5 | 7 | 1·99 |
| 3·0 | 3 | 7 | 9 | 2·77 |
| 2·6 | 5 | 11 | 13 | 4·34 |
| 2·3 | 6 | 13 | 16 | 5·51 |
| 2·02 | 9 | 17 | 23 | 8·38 |
| 1·82 | 13 | 23 | 32 | 12·19 |
| 1·61 | 21 | 34 | 50 | 20·04 |
| 1·50 | 30 | 45 | 69 | 28·53 |
| 1·336 | 63 | 83 | 138 | 60·31 |

*Source*: By permission, from *Industrial Statistics*, by Paul Peach, copyright 1947, Edwards and Broughton Company, Raleigh, North Carolina.

## AVERAGE RANGE METHOD DIVISORS FOR ESTIMATING THE STANDARD DEVIATION OF ACCOUNTING POPULATIONS

After the average range has been calculated the following divisors are used to arrive at the estimate of standard deviation.

Table 21.6. Divisors for estimating standard deviation.

| Group size | Divisor |
|---|---|
| 6 | 2·5344 |
| 7 | 2·7043 |
| 8 | 2·8472 |
| 9 | 2·9700 |
| 10 | 3·0775 |

Table 21.7.1. A table of general random numbers.

| Col. / Line | (1) | (2) | (3) | (4) | (5) | (6) |
|---|---|---|---|---|---|---|
| 751 | 03548 | 52011 | 53722 | 62927 | 01693 | 90948 |
| 752 | 05066 | 74263 | 11659 | 84658 | 52063 | 42299 |
| 753 | 19814 | 89956 | 36256 | 19896 | 56654 | 69424 |
| 754 | 44928 | 11206 | 21377 | 35086 | 62233 | 93761 |
| 755 | 52158 | 08923 | 94812 | 04443 | 32028 | 96465 |
| 756 | 53211 | 58616 | 05135 | 52204 | 51079 | 11341 |
| 757 | 27334 | 43976 | 14685 | 38119 | 29486 | 57290 |
| 758 | 66166 | 06709 | 69495 | 42150 | 38018 | 43875 |
| 759 | 01837 | 16750 | 96491 | 33095 | 30383 | 48804 |
| 760 | 38295 | 51093 | 63495 | 13203 | 93562 | 94132 |
| 761 | 72510 | 80830 | 39948 | 94133 | 00780 | 14167 |
| 762 | 16354 | 00336 | 06494 | 30078 | 46134 | 62486 |
| 763 | 45168 | 70700 | 04592 | 35281 | 47737 | 28881 |
| 764 | 68137 | 13619 | 84666 | 78104 | 83546 | 72551 |
| 765 | 31848 | 95753 | 76858 | 89517 | 91138 | 62356 |
| 766 | 17216 | 37292 | 05495 | 50885 | 98994 | 32966 |
| 767 | 73211 | 42922 | 57386 | 95490 | 56100 | 08977 |
| 768 | 29217 | 56753 | 90171 | 87554 | 57421 | 35830 |
| 769 | 66780 | 34571 | 71684 | 28798 | 47123 | 25232 |
| 770 | 55780 | 40081 | 93674 | 70837 | 92534 | 65892 |
| 771 | 05788 | 64237 | 58140 | 63870 | 60170 | 17829 |
| 772 | 21911 | 51065 | 51526 | 65122 | 52608 | 52836 |
| 773 | 32645 | 38861 | 25181 | 16042 | 31903 | 46525 |
| 774 | 72304 | 15382 | 01151 | 63162 | 23656 | 69649 |
| 775 | 27637 | 04122 | 86132 | 22538 | 98976 | 90718 |
| 776 | 66305 | 85906 | 87925 | 50081 | 37585 | 63674 |
| 777 | 53470 | 40332 | 81044 | 00558 | 50403 | 48029 |
| 778 | 01355 | 77096 | 64828 | 02445 | 94511 | 09503 |
| 779 | 66954 | 04728 | 86153 | 10933 | 86557 | 10877 |
| 780 | 53734 | 15628 | 08080 | 24011 | 04187 | 65722 |
| 781 | 23114 | 08743 | 07186 | 23825 | 04298 | 72839 |
| 782 | 16803 | 91335 | 64192 | 13631 | 20338 | 74858 |
| 783 | 57833 | 99034 | 38028 | 32038 | 81270 | 77809 |
| 784 | 83203 | 59665 | 72314 | 67942 | 01380 | 84467 |
| 785 | 79299 | 97340 | 04568 | 65386 | 04876 | 31514 |
| 786 | 99442 | 63865 | 14360 | 83898 | 98873 | 17471 |
| 787 | 77773 | 19621 | 81557 | 84629 | 18808 | 89056 |
| 788 | 18581 | 12572 | 15185 | 57989 | 87644 | 88902 |
| 789 | 80978 | 13327 | 19682 | 53353 | 96223 | 04775 |
| 790 | 96884 | 95522 | 04791 | 93463 | 20316 | 84054 |
| 791 | 61113 | 82511 | 44196 | 73740 | 16111 | 73200 |
| 792 | 01272 | 75657 | 28365 | 17431 | 93603 | 32457 |
| 793 | 82357 | 77572 | 75628 | 93073 | 38281 | 72103 |
| 794 | 20434 | 70899 | 44243 | 19741 | 59954 | 73617 |
| 795 | 38580 | 10399 | 44894 | 25476 | 04984 | 64543 |
| 796 | 20211 | 12980 | 45861 | 82587 | 87534 | 39405 |
| 797 | 98954 | 03124 | 09433 | 19894 | 01380 | 18962 |
| 798 | 99749 | 02140 | 46641 | 56354 | 78746 | 89410 |
| 799 | 47167 | 24716 | 84417 | 40097 | 46608 | 11667 |
| 800 | 65812 | 77947 | 27864 | 98144 | 01818 | 28214 |

Table 21.7.1 (*continued*)

| (7) | (8) | (9) | (10) | (11) | (12) | (13) | (14) |
|---|---|---|---|---|---|---|---|
| 75340 | 16660 | 00939 | 77148 | 53778 | 20615 | 97851 | 58353 |
| 94340 | 20391 | 00080 | 43359 | 44231 | 06891 | 86588 | 88565 |
| 84446 | 95294 | 00919 | 60267 | 34349 | 64353 | 92469 | 01606 |
| 56000 | 42066 | 35898 | 48944 | 36158 | 33177 | 16239 | 76624 |
| 94430 | 42834 | 22836 | 88816 | 64467 | 18329 | 30144 | 31536 |
| 20366 | 86248 | 35160 | 31465 | 17985 | 69726 | 18722 | 52722 |
| 34141 | 18058 | 91299 | 58001 | 17944 | 00296 | 13949 | 94904 |
| 44534 | 48712 | 98089 | 76889 | 55780 | 09038 | 95667 | 57007 |
| 21228 | 10863 | 08350 | 25610 | 80866 | 00115 | 27666 | 94607 |
| 54810 | 11410 | 73776 | 55752 | 36158 | 52297 | 31528 | 77790 |
| 83322 | 32747 | 97291 | 66126 | 64996 | 21354 | 64615 | 64234 |
| 70923 | 17400 | 28797 | 83599 | 78655 | 67488 | 09715 | 69232 |
| 57221 | 79935 | 99756 | 61564 | 36936 | 55668 | 87148 | 95939 |
| 40848 | 52138 | 36343 | 93975 | 09556 | 58888 | 08125 | 80961 |
| 66405 | 98171 | 66239 | 49761 | 88800 | 29069 | 30175 | 31267 |
| 83496 | 54614 | 75214 | 34959 | 15500 | 32107 | 81638 | 46696 |
| 78499 | 53540 | 41745 | 09805 | 67110 | 53329 | 92776 | 32312 |
| 60961 | 87828 | 84777 | 19688 | 42464 | 35569 | 84904 | 13130 |
| 67995 | 39568 | 35855 | 61645 | 64448 | 65170 | 46792 | 77081 |
| 32422 | 02887 | 03170 | 05372 | 48068 | 65758 | 62376 | 94055 |
| 69038 | 02500 | 21071 | 26294 | 95005 | 47815 | 87991 | 93358 |
| 86157 | 19943 | 63173 | 34921 | 11943 | 36931 | 74722 | 06193 |
| 96290 | 24323 | 68269 | 45841 | 73918 | 13720 | 52336 | 48416 |
| 54580 | 33479 | 62899 | 33716 | 53668 | 86735 | 36746 | 21939 |
| 81840 | 33461 | 28526 | 96231 | 28082 | 89710 | 16038 | 26648 |
| 65144 | 75814 | 87596 | 04642 | 07590 | 82276 | 59336 | 84262 |
| 07719 | 83340 | 20258 | 49737 | 45499 | 72241 | 90103 | 96141 |
| 12283 | 15264 | 32245 | 55594 | 03668 | 05664 | 67999 | 06001 |
| 97049 | 40124 | 87071 | 03864 | 29526 | 42888 | 09057 | 82966 |
| 82863 | 59616 | 60664 | 76761 | 84417 | 72560 | 77869 | 42603 |
| 43963 | 52136 | 60923 | 88568 | 55458 | 87333 | 81939 | 67361 |
| 67813 | 68131 | 56695 | 93253 | 71650 | 81735 | 13302 | 30319 |
| 99813 | 67091 | 56997 | 62005 | 95516 | 34011 | 52035 | 79500 |
| 27813 | 74783 | 56696 | 84101 | 93859 | 12016 | 20869 | 57009 |
| 78813 | 77887 | 66697 | 51048 | 28395 | 62626 | 42911 | 99130 |
| 01813 | 81541 | 66913 | 52874 | 10563 | 59450 | 16885 | 16009 |
| 48194 | 65899 | 86244 | 74431 | 24232 | 33299 | 58332 | 18314 |
| 49946 | 76235 | 92633 | 10232 | 84241 | 66467 | 86018 | 92272 |
| 05590 | 55237 | 81695 | 31870 | 48893 | 94058 | 88943 | 77040 |
| 68980 | 56907 | 77188 | 54803 | 67622 | 57968 | 78532 | 66688 |
| 65980 | 66320 | 77327 | 14715 | 24759 | 08145 | 28077 | 32303 |
| 57980 | 66157 | 87797 | 55887 | 57788 | 29337 | 73409 | 35682 |
| 39980 | 71150 | 87282 | 01585 | 25850 | 46138 | 21777 | 97555 |
| 11661 | 81574 | 05903 | 25909 | 38638 | 32196 | 41345 | 22152 |
| 14141 | 97911 | 89418 | 97980 | 49571 | 60483 | 94714 | 51753 |
| 03701 | 94381 | 83777 | 61546 | 81541 | 69603 | 61768 | 40328 |
| 40701 | 05185 | 23280 | 04945 | 49801 | 98051 | 77848 | 62320 |
| 35701 | 00250 | 23508 | 51246 | 09460 | 02313 | 71279 | 74223 |
| 40701 | 06155 | 23784 | 68961 | 34289 | 04531 | 12752 | 08516 |
| 83701 | 10257 | 33945 | 08795 | 30976 | 14017 | 59161 | 31388 |

Table 21.7.2. A table of general random numbers.

| Line \ Col. | (1) | (2) | (3) | (4) | (5) | (6) |
|---|---|---|---|---|---|---|
| 801 | 33993 | 51249 | 78123 | 16507 | 57399 | 77922 |
| 802 | 39041 | 05779 | 74278 | 75301 | 01779 | 60768 |
| 803 | 56011 | 26839 | 38501 | 03321 | 43259 | 73148 |
| 804 | 07397 | 95853 | 45764 | 43803 | 76659 | 57736 |
| 805 | 74998 | 53337 | 13860 | 89430 | 95825 | 65893 |
| 806 | 59572 | 95893 | 69765 | 43597 | 90570 | 60909 |
| 807 | 74645 | 13940 | 28640 | 00127 | 04261 | 17650 |
| 808 | 42765 | 23855 | 38451 | 11462 | 32671 | 52126 |
| 809 | 66561 | 56130 | 30356 | 54034 | 53996 | 98874 |
| 810 | 50670 | 13172 | 31460 | 20224 | 34293 | 59458 |
| 811 | 53971 | 08701 | 38356 | 36149 | 10891 | 05178 |
| 812 | 47177 | 03085 | 37432 | 94053 | 87057 | 61859 |
| 813 | 41494 | 89270 | 48063 | 12253 | 00383 | 96010 |
| 814 | 07409 | 32874 | 03514 | 84943 | 74421 | 86708 |
| 815 | 03097 | 12212 | 43093 | 46224 | 14431 | 15065 |
| 816 | 34722 | 88896 | 59205 | 18004 | 96431 | 41366 |
| 817 | 48117 | 83879 | 52509 | 29339 | 87735 | 97499 |
| 818 | 14628 | 89161 | 66972 | 19180 | 40852 | 91738 |
| 819 | 61512 | 79376 | 88184 | 29415 | 50716 | 93393 |
| 820 | 99954 | 55656 | 01946 | 57035 | 64418 | 29700 |
| 821 | 61455 | 28229 | 82511 | 11622 | 60786 | 18442 |
| 822 | 10398 | 50239 | 70191 | 37585 | 98373 | 04651 |
| 823 | 59075 | 81492 | 40669 | 16391 | 12148 | 38538 |
| 824 | 91497 | 76797 | 82557 | 55301 | 61570 | 69577 |
| 825 | 74619 | 62316 | 80041 | 53053 | 81252 | 32739 |
| 826 | 12536 | 80792 | 44581 | 12616 | 49740 | 86946 |
| 827 | 10246 | 49556 | 07610 | 59950 | 34387 | 70013 |
| 828 | 92506 | 24397 | 19145 | 24185 | 24479 | 70118 |
| 829 | 65745 | 27223 | 22831 | 39446 | 65808 | 95534 |
| 830 | 01707 | 04494 | 48168 | 58480 | 74983 | 63091 |
| 831 | 66959 | 80109 | 88908 | 38759 | 80716 | 36340 |
| 832 | 79278 | 02746 | 50718 | 90196 | 28394 | 82035 |
| 833 | 11343 | 22312 | 41379 | 22297 | 71703 | 78729 |
| 834 | 40415 | 10553 | 65932 | 34938 | 43938 | 39262 |
| 835 | 72774 | 25480 | 30264 | 08291 | 93796 | 22281 |
| 836 | 75886 | 86543 | 47020 | 14493 | 38363 | 64238 |
| 837 | 64628 | 20234 | 07967 | 46676 | 42907 | 60909 |
| 838 | 45905 | 77701 | 98976 | 70056 | 80502 | 68650 |
| 839 | 77691 | 00408 | 64191 | 11006 | 39212 | 26862 |
| 840 | 39172 | 12824 | 43379 | 57590 | 45307 | 72206 |
| 841 | 67120 | 01558 | 99762 | 79752 | 17139 | 52265 |
| 842 | 88264 | 85390 | 92841 | 63811 | 64423 | 50910 |
| 843 | 78097 | 59495 | 45090 | 74592 | 47474 | 56157 |
| 844 | 41888 | 69798 | 82296 | 09312 | 04150 | 07616 |
| 845 | 46618 | 07254 | 28714 | 18244 | 53214 | 39560 |
| 846 | 29213 | 42101 | 25089 | 11881 | 77558 | 72738 |
| 847 | 38601 | 25735 | 04726 | 36544 | 67842 | 93937 |
| 848 | 92207 | 10011 | 64210 | 77096 | 00011 | 79218 |
| 849 | 30610 | 13236 | 33241 | 68731 | 30955 | 40587 |
| 850 | 74544 | 72506 | 62226 | 65685 | 37996 | 00377 |

Table 21.7.2. (*continued*)

| (7) | (8) | (9) | (10) | (11) | (12) | (13) | (14) |
|---|---|---|---|---|---|---|---|
| 38198 | 63494 | 00278 | 30782 | 33119 | 64943 | 17239 | 69020 |
| 22023 | 07510 | 67883 | 55288 | 67391 | 54188 | 31913 | 29733 |
| 43615 | 49093 | 91641 | 77179 | 50837 | 48734 | 85187 | 41210 |
| 44801 | 45623 | 23714 | 69657 | 87971 | 24757 | 94493 | 78723 |
| 96572 | 73975 | 19577 | 87947 | 23962 | 78235 | 64839 | 73456 |
| 06478 | 77692 | 30911 | 08272 | 81887 | 57749 | 02952 | 51524 |
| 34050 | 78788 | 57948 | 36189 | 88382 | 72324 | 59253 | 30258 |
| 23800 | 02691 | 57034 | 34532 | 19711 | 71567 | 90495 | 55980 |
| 78001 | 29707 | 91938 | 72016 | 16429 | 69726 | 41990 | 33673 |
| 24410 | 01366 | 68825 | 22798 | 52873 | 18370 | 15577 | 63271 |
| 55653 | 31553 | 20037 | 39346 | 28591 | 13505 | 04446 | 92130 |
| 97943 | 81113 | 62161 | 11369 | 54419 | 58886 | 89956 | 12857 |
| 41457 | 54657 | 46881 | 75255 | 29242 | 07537 | 53186 | 95083 |
| 34267 | 66071 | 62262 | 99391 | 61245 | 95839 | 75203 | 93984 |
| 18267 | 60039 | 62089 | 38572 | 70988 | 17279 | 05469 | 28591 |
| 50982 | 92400 | 59369 | 43605 | 26404 | 04176 | 05106 | 08366 |
| 42848 | 81449 | 80024 | 81312 | 59469 | 91169 | 70851 | 90165 |
| 23920 | 75518 | 32041 | 13411 | 61334 | 52386 | 33582 | 72143 |
| 96220 | 82277 | 64510 | 43374 | 09107 | 28813 | 41848 | 08813 |
| 99242 | 42586 | 11583 | 82768 | 44966 | 39192 | 82144 | 05810 |
| 36508 | 98936 | 19050 | 57242 | 33045 | 54278 | 21720 | 87812 |
| 67804 | 84062 | 27380 | 75486 | 63171 | 24529 | 60070 | 66939 |
| 73873 | 68596 | 25538 | 83646 | 61066 | 45210 | 24182 | 18687 |
| 23301 | 31921 | 09862 | 73089 | 69329 | 41916 | 41165 | 34503 |
| 65201 | 92165 | 93792 | 30912 | 59105 | 76944 | 70998 | 00317 |
| 41819 | 85104 | 25705 | 92481 | 95287 | 61769 | 29390 | 05764 |
| 64460 | 96719 | 43056 | 24268 | 23303 | 19863 | 43644 | 76986 |
| 42708 | 54311 | 95989 | 08402 | 77608 | 98356 | 47034 | 01635 |
| 03348 | 11435 | 24166 | 62726 | 99878 | 59302 | 81164 | 08010 |
| 81027 | 72579 | 67249 | 48089 | 34219 | 71727 | 86665 | 94975 |
| 30082 | 43295 | 37551 | 18531 | 43903 | 94975 | 31049 | 19033 |
| 03255 | 39574 | 41483 | 12450 | 32494 | 65192 | 54772 | 97431 |
| 65082 | 57759 | 79579 | 41516 | 46248 | 37348 | 34631 | 88164 |
| 95828 | 98617 | 27401 | 50226 | 17322 | 44024 | 23133 | 57899 |
| 51434 | 66771 | 20118 | 00502 | 07738 | 31841 | 90200 | 46348 |
| 16322 | 45503 | 90723 | 35607 | 43715 | 85751 | 15888 | 80645 |
| 73293 | 38588 | 31035 | 12226 | 37746 | 45008 | 43271 | 32015 |
| 24469 | 15574 | 40018 | 90057 | 96540 | 47174 | 03943 | 37553 |
| 99863 | 58155 | 66052 | 96864 | 61790 | 11064 | 49308 | 94510 |
| 53283 | 75882 | 93431 | 44830 | 06300 | 45456 | 49567 | 51673 |
| 97997 | 66806 | 55559 | 62043 | 51384 | 32423 | 88325 | 99634 |
| 38189 | 88183 | 56625 | 22910 | 58230 | 70491 | 71111 | 37202 |
| 88287 | 47032 | 66341 | 38328 | 70538 | 91105 | 12056 | 36125 |
| 34572 | 83202 | 58691 | 27354 | 37015 | 11278 | 49697 | 65667 |
| 68753 | 16825 | 48639 | 38228 | 95166 | 53649 | 05071 | 26894 |
| 57234 | 28458 | 74313 | 29665 | 97366 | 94714 | 48704 | 07033 |
| 68745 | 62979 | 97750 | 28293 | 75891 | 08362 | 71546 | 17993 |
| 52123 | 29841 | 76145 | 82364 | 55774 | 15462 | 44555 | 26844 |
| 45206 | 11949 | 28295 | 12666 | 98479 | 82498 | 49195 | 46254 |
| 59917 | 91100 | 07993 | 15046 | 51303 | 19515 | 25055 | 56386 |

Table 21.8. Finite population correction factor.

| Percentage of population in sample $n/N$ % | Finite population correction factor |
|---|---|
| 1 | 0·995 |
| 2 | 0·990 |
| 3 | 0·985 |
| 4 | 0·980 |
| 5 | 0·975 |
| 6 | 0·970 |
| 7 | 0·964 |
| 8 | 0·959 |
| 9 | 0·954 |
| 10 | 0·949 |
| 11 | 0·943 |
| 12 | 0·938 |
| 13 | 0·933 |
| 14 | 0·927 |
| 15 | 0·922 |
| 16 | 0·917 |
| 17 | 0·911 |
| 18 | 0·906 |
| 19 | 0·900 |
| 20 | 0·894 |
| 21 | 0·889 |
| 22 | 0·883 |
| 23 | 0·878 |
| 24 | 0·872 |
| 25 | 0·866 |
| 26 | 0·860 |
| 27 | 0·854 |
| 28 | 0·849 |
| 29 | 0·843 |
| 30 | 0·837 |
| 31 | 0·831 |
| 32 | 0·825 |
| 33 | 0·819 |
| 34 | 0·812 |
| 35 | 0·806 |
| 36 | 0·800 |
| 37 | 0·794 |
| 38 | 0·787 |
| 39 | 0·781 |
| 40 | 0·775 |
| 45 | 0·742 |
| 50 | 0·712 |

## RANDOM DAYS

Table 21.9. Table of random days.

| | |
|---|---|
| Sunday | 1 |
| Monday | 2 |
| Tuesday | 3 |
| Wednesday | 4 |
| Thursday | 5 |
| Friday | 6 |
| Saturday | 7 |

| | | | | | | | | | | | | | | | | | | | |
|---|---|---|---|---|---|---|---|---|---|---|---|---|---|---|---|---|---|---|---|
| 1 | 4 | 1 | 6 | 4 | 7 | 4 | 5 | 7 | 4 | 6 | 4 | 3 | 4 | 4 | 2 | 4 | 5 | 4 | 2 |
| 5 | 7 | 1 | 7 | 3 | 5 | 4 | 1 | 3 | 7 | 1 | 2 | 2 | 3 | 5 | 1 | 1 | 2 | 7 | 3 |
| 3 | 2 | 3 | 1 | 3 | 7 | 6 | 2 | 1 | 7 | 4 | 3 | 3 | 3 | 5 | 3 | 7 | 7 | 4 | 3 |
| 5 | 1 | 3 | 6 | 4 | 1 | 4 | 2 | 5 | 4 | 1 | 2 | 6 | 6 | 4 | 2 | 6 | 4 | 1 | 4 |
| 6 | 2 | 3 | 5 | 3 | 1 | 6 | 3 | 1 | 6 | 3 | 6 | 2 | 5 | 7 | 4 | 4 | 2 | 5 | 6 |
| 7 | 6 | 7 | 6 | 1 | 1 | 2 | 4 | 3 | 6 | 3 | 7 | 2 | 4 | 4 | 7 | 3 | 4 | 4 | 7 |

## RANDOM WEEKS

Table 21.10. Table of random weeks. Start at any point in the table and move in any direction to select a series of random weeks.

| | | | | |
|---|---|---|---|---|
| 03 | 49 | 41 | 51 | 16 |
| 23 | 07 | 32 | 01 | 50 |
| 03 | 38 | 24 | 11 | 31 |
| 47 | 51 | 37 | 07 | 40 |
| 52 | 38 | 13 | 10 | 05 |
| 41 | 19 | 44 | 30 | 44 |
| 07 | 30 | 27 | 05 | 23 |
| 31 | 41 | 47 | 05 | 22 |
| 50 | 31 | 41 | 16 | 13 |
| 33 | 46 | 34 | 18 | 21 |

## RANDOM MONTHS

Table 21.11. Table of random months.

| | | | | |
|---|---|---|---|---|
| 11 | 07 | 10 | 09 | 07 |
| 06 | 08 | 03 | 11 | 04 |
| 12 | 09 | 01 | 08 | 03 |
| 08 | 01 | 10 | 01 | 07 |
| 02 | 06 | 12 | 05 | 02 |
| 03 | 03 | 01 | 12 | 05 |
| 10 | 04 | 11 | 02 | 06 |
| 07 | 02 | 05 | 09 | 03 |
| 08 | 05 | 09 | 08 | 03 |
| 12 | 05 | 08 | 03 | 09 |

# References

AICPA (1973). *Statement on Auditing Standards*, AICPA Publications, New York.

Aly, H. F., and Duboff, J. I. (1971). 'Statistical v. judgement sampling', *Accounting Review*, **46**, 1, 119–128.

Anderson, R., and Teitlebaum, A. (1973). 'Dollar unit sampling', *Canadian Chartered Accountant*, April.

Arkin, H. (1963). *Handbook of Sampling for Auditing and Accounting. Vol. 1, Methods*, McGraw–Hill, New York.

Brown, R. G., and Vance, L. L. (1961). *Sampling Tables for Estimating Error Rates or Other Proportions*, Institute of Business and Economic Research, University of California, Berkeley, California.

Cochran, W. G. (1963). *Sampling Techniques*, John Wiley, New York.

Committee on Statistical Sampling (1962). 'Statistical sampling and the independent auditors', *Journal of Accountancy*, **113** (February 1962), 61.

Cyert, R. M., and Trueblood, R. M. (1957). 'Statistical sampling techniques in the aging of accounts receivable in a department store', *Management Science*, **3**, 2, 185–195.

Dalleck, W. C. (1960). 'Inductive accounting—an application of statistical sampling techniques', *O.R. Applied Science Report*, No. 17, American Management Association, New York.

Davidson, H. J. (1959). 'Accuracy in statistical sampling', *Accounting Review*, **34**, 356–365.

Davis, G. B., Neter, J., and Palmer, R. R. (1967). 'An experimental study of audit confirmations', *Journal of Accountancy*, **123** (June 1967), 36–40.

Dodge, H. F., and Romig, H. G. (1944). *Sampling Inspection Tables*, John Wiley, New York.

Galton, F. (1907). *Memories*, Methuen, London, 251.

Grubbs, F. E., and Weaver, C. L. (1947). 'The best unbiased estimate of the population standard deviation based on group ranges', *Journal of the American Statistical Association*, **42**, 224–241.

Hall, W. D. (1967). 'Inventory determination by means of statistical sampling when clients have perpetual records', *Journal of Accountancy*, **123** (March 1967), 65–71.

Haworth, T. G. (1969). 'Statistical sampling: a practical approach', *Accountancy*, **80** (February 1969), 101–109.

Hoel, P. G. (1960). *Elementary Statistics*, John Wiley, New York.

Institute of Internal Auditors (1970). *Sampling Manual for Auditors*, New York.

Irvine, J. R (1964). 'Case study on the use of computer and statistical techniques', *Journal of Accountancy*, **117** (April 1964), 67–68.

Kraft, W. H. (1968). 'Statistical sampling for auditors. A new look', *Journal of Accountancy*, **126** (August 1968), 49–56.

London Commissioner of Police (1973). *Report of London Commissioner of Police*, UK Government Printing Office.

Meikle, G. R. (1972). *Statistical Sampling in an Audit Context*, Canadian Institute of Chartered Accountants.

Rudell, A. (1957). 'Applied sampling doubles inventory accuracy, halves cost', *N.A.A. Bulletin*, **39**, 2 (October 1957), 5–11.

Smurthwaite, J. (1965). 'Statistical sampling techniques as an audit tool', *Accountancy*, **76**, 859 (March 1965), 201–209.

Stringer, K. (1963). 'Practical aspects of statistical sampling in auditing', *Proceedings of the American Statistical Association*, 405–411.

Tracy, J. A. (1969). 'Bayesian statistical methods in auditing', *Accounting Review*, **44**, 3, 555–561.

Trentin, H. G. (1968). 'Sampling in auditing—a case study', *Journal of Accountancy*, **125** (March 1968), 39–43.

U.S. Air Force (1960). *Tables of Probabilities for Use in Stop or Go Sampling*, Government Printing Office, Washington D.C.

U.S. Department of Defence (1950). *Sampling Procedures and Tables for Inspection by Attributes*, Government Printing Office, Washington D.C.

Vanasse, R. W. (1968). *Statistical Sampling for Auditing and Accounting Decisions: A Simulation*, McGraw–Hill, New York.

Vance, L. L. (1960). 'Review of developments in statistical sampling for accountants', *Accounting Review*, **35** (January 1960), 19–28.

Vance, L. L., and Neter, J. (1956). *Statistical Sampling for Auditors and Accountants*, John Wiley, New York.

# Author Index

# Subject Index